HOLONOMY

holonomy, holo′•nomy n. (Gr. *holos*, entire, complete, whole and *nomos*, a law or rule.) The science or systematic study of wholes, of entire systems, etc. Basic to holonomy is the simultaneous interrelatedness and interpenetration of all phenomena. Basic to holonomy is the assumption of unity and oneness as opposed to fragmentation, isolation, and separateness.

Library of Congress Cataloging in Publication Data

Samples, Bob.
Mind of our mother.

Includes index.
1. Holism. 2. Mind and body. 3. Personality and culture. 4. Man—Influence of environment. I. Title.
BF204.S25 150.19 81-12822
AACR2
ISBN 0-201-06668-8 (pbk.)

Printed in the United States of America. Published simultaneously in Canada. Library of Congress Catalog Card No. 81-12822.

All photographs and drawings by Bob Samples, except those which are by Olina Gilbert, on pages xiv, 42, 50, 78, 105, 135.
Design and graphics by Bob Samples. Cover design by Bob Samples.
Cover photographs courtesy of Mount Palomar and Mount Wilson Observatories and the National Aeronautics and Space Administration.

Typesetting by Graphic Directions, Boulder, Colorado.

ISBN 0-201-06668-8

ABCDEFGHIJ-HA-8987654321

First printing, September 1981

MIND of our MOTHER

This book is for the Mothers of my life . . .

Eve who was Mother to me
Cheryl who is Mother to Stician
Olina who is Mother to Mark,
Michael and Glen

CONTENTS

SPIRIT

I observe purpose in the universe. I see absolutely nothing that is without purpose. From the smallest particle of an atom to the universe itself, it is full of purpose. There is seemingly intelligence behind it. It *works*—on a time scale that perhaps makes our day-to-day existence in this world insignificant.

To me, ours is a living, breathing universe. It contains life. Everywhere I see life . . . there is purpose, from the smallest blade of grass to you and me. As I look at things, it is pretty easy to figure out purpose. I am not talking about reason . . . but purpose. There is a purpose in the existence of all things. And yet so many of us seem to question our own purpose.

Human experience is the same for all of us. We are human beings and we experience the world. It is most clearly that which binds us together . . . the spirit of life itself. It is time that we all start representing this spirit in each and every thing that we do. It is time to make the world work.

The world was made to work . . . but it hasn't been working too well lately. If we pull it out, it's going to be because we choose to do it. We will start acting again like human beings.

I know that we have everything it takes to have life grow fruitfully, multiply, and be fulfilling and satisfying for every living creature. we have all it takes. I believe that we are here *for* each other, not against each other. We are here for each other, and for the world itself. It is when we start acting in this spirit we will make the world work.

Unfortunately, we live in a world of separationist tendencies. Everyone is supporting some kind of separation—some people by not thinking about it, not doing anything, not caring. Other people support it consciously by encouraging us to be against each other.

I think that we have come to a time where our survival as human beings requires us to totally turn around. I am talking about a shift in consciousness. I want to propose to you that we turn it around . . . . What I am here for is simply that. I wish to encourage our honoring of harmony in the universe, harmony between people. I want us to be living, breathing, as a single organism . . . committed to a life that is worth living.

John Denver

I am looking for my family
And all of you are mine.
We are standing all together
Face to face and arm in arm.
We are standing on the threshold of a dream.

No more hunger, no more killing,
No more wasting life away;
It is simply an idea
And I know its time has come.

I want to live, I want to grow,
I want to see, I want to know,
I want to share what I can give.
I want to be
I want to live . . .
I want to live!

THE JOURNEY

The mother referred to in the title of this book is Mother Earth. When I was a child on the plains edging the Colorado front range, I used to climb up into elm trees on hot summer nights when thunderstorms boiled up out of the sugar beet fields to the east. The wind whipped the trees into a passionate dance, and I would rock back and forth to rhythms I loved but could not explain. My soul was sculptured by the flashes of lightning, the roll of thunder, the surprising coldness of the rain and the shiver that followed it back into the air as it evaporated off my skin.

Other times I would walk in arroyos choked with tumble weeds and rounded pebbles, smoothed by rivers now gone to other valleys. I loved the earth even though I didn't know it was there. Later I learned it was a planet, the third planet from the sun. I learned that storms had washed those pebbles away from glaciers, and that the rock had been born hundreds of millions of years before it was smoothed into shapes I could toss at imaginary jack rabbits. I learned of igneous, sedimentary, and metamorphic. I learned two college degrees worth of earth—so much did I learn that my childhood love had faded into an adult kind of purpose. It had become more important to gather knowledge than to wander in mysteries.

My first job in geology and geophysics cured me quickly of my knowledge-filled arrogance. Deep holes were drilled, slurry-like gray blood seeped out and away from the drill sites. Explosives were shoved deep in the cavities and great unseen holes were blasted a hundred feet below us. Geophones listened to the trauma. They passed the shockwaves to a truck where intricate roles of squiggles were captured and studied by eyes narrowed with purpose. The purpose was to exploit, to achieve gain . . . to judge the worth of the earth. Almost at once and without warning my childhood came back. I realized that I had studied geology, geophysics, and astronomy because I had been in love with the earth. I stood shocked, like a medical student at his or her first autopsy, as I looked upon the neatness of the drill hole wounds. I loved the earth and I had become a rapist.

My knowledge, my wisdom, my training prepared me to tear the earth apart. To rip from the mother's body fetal enclosures that sated the greed of some invisible and distant ledger. My first career lasted ten weeks. I became an ex-geologist. Could I really be an ex-anything? Could I erase the joy I had felt *before* the autopsy? Could I forget the way I felt when I found my first fossils, and the sense of "fit" that came when I traced their citation in *Index Fossils of North America*?

No, I could not. My years of knowledge were not my enemy . . . they were not a cloak of evil. That which was evil was the blind willingness I exhibited when I began to use this knowledge to exploit rather than to celebrate. For the next twenty-five years, I explored how to celebrate what I knew, how I knew, and how all related to the earth . . . my true first-love.

This book is about that journey. It is told in the language of the West. I am a Westerner trained in Western tradition. I love logic, analytic thought, and the tidiness of an equation that gracefully says some tiny thing about the way the world works. I love these things, but I do not worship them.

My exploration of planetary consciousness has avoided the meadows of esoteric traditions. In this book, there is little of the tendency of many colleagues who rush to Eastern traditions to immerse themselves in a supposed antidote to Western dogma. I have found Eastern traditions as dogmatic. For me the samurai is as much a mindless bigot as our lonely frontier gunfighter. There is no significant history of honoring the planet in either Eastern or Western traditions. China had been de–forested, eroded, and savaged at least eight times before Marco Polo arrived.

In human terms, my love for the earth is most honored by native Indian world views. These are expressions of similar world views held by traditional people whose life technologies are fundamentally non-exploitative, who until recent times lived in equilibrium with the planet's offerings, who mostly represent a Southern hemisphere world view. I see the Native American Indian as a northern incursion of this consciousness. I see Eastern and Western traditions as complementary versions of an exploitative Northern hemisphere world view.

This book is *not* a "do-it-yourself" planetary consciousness checklist. There are no slick technologies to spirituality . . . and planetary consciousness is a spiritual issue. I explore the source of spirituality in humans—the mind—and how it can blend into the spirituality of the planet. I say much about how humans have generally divested themselves of their planetary bonding. But I also cite the evidence for—and the nature of—the evolutionary transformation I believe to be taking place.

Dozen of my colleagues trace this journey in other ways. In one sense, I see *Mind of Our Mother* as the third book in a trilogy written independently and from different perspectives. Each of these separate, non-collaborative efforts is part of the story of this evolving consciousness. The first was Theodore Roszak's *Person/Planet*; the second, Marilyn Ferguson's *Aquarian Conspiracy*.

Person/Planet looks fully into the face of the "creative" disintegration of industrial society. *Aquarian Conspiracy* chronicles the specific expression of the vision, biases, and methodologies of those who are engaged in this "creative" transformation. *Mind of Our Mother* allies itself with both of these works, but in a different way. This book attempts to state from a perspective of consciousness the process of reunion of personal, cultural, and natural harmonies.

Bob Samples

HAWKSONG
in the front range
of the Rocky Mountains,
Colorado.

INTRODUCTION

Bob Samples' book, *Mind of our Mother*, is the best handbook I have yet seen for elevating our human awareness—our consciousness—to the whole that surrounds us. The whole that surrounds us is our mother planet . . . the earth. Below us is the earth's solid state, beside us the liquid state, and above us the gaseous state. Each is a domain of our mother planet.

If and until we can use ourselves in the service of the land, seas, and atmosphere in an holonomous way, we may end up in a quantum leap into total destruction.

Each of us is a center of the universe extending outward from us. This extension is an holonomous reality to which we belong and from which we have evolved. We do not know where we are going, nor do we know what happens to our consciousness when this vehicle in which we are traveling dies. Perhaps we become, as Bob feels, a part of the whole. We rejoin the holonomous nature of the land upon which we now walk, the waters in which we now swim, and the atmosphere which we now breathe.

Our playmates, the dolphins, live their whole lives in the sea. They have lived on this planet for fifteen million years, with essentially the same size brains that they have now. Out of the fifteen billion years the earth has been here, dolphins have a much longer track record than humans. Consider the time scale: fifteen billion years ago the planet was created, three billion years ago life appeared in the oceans, fifteen million years ago the dolphins appeared . . . and only 300,000 years ago humans achieved the brain size of dolphins.

Therefore, to really appreciate holonomy—the mind of our mother—we should establish communication with the dolphins and the whales. I am sure they know the holonomous mind of mother earth. They can teach us to achieve that unity which Bob so beautifully presents in this book.

I have found *Mind of Our Mother* to be a creative handbook which relates me to many of the things which I have explored in the past. In fact, these are no longer "things"—they are holonomies. They are intertwined and interrelated, and the unitary field theory eventually will come out of such views. I just hope that when the unitary field theory arrives in enough people's minds, it will not be treated as a weapon of war. Instead, I hope it will be treated as a gift of understanding—one that unites us with the mind of our mother.

The mind of our mother is central to all of us—the dolphins, the whales, our children, and the planet. I hope that thousands of generations after us will join in celebration of mother earth and for the first time become their birthright.

John C. Lilly, M.D.

We have in the last 20 years been exposed to a vast amount of information about self. Freudian psychology expanded the concept of self, or mind inwards to include the whole system of communication within the body—the automatic, the habitual and the vast range of unconscious process. With some of us the Freudian was not enough and we explored the esoteric. Every kind of imported mystical trip was at least politely tolerated by Americans everywhere. We have watched our children join any and every kind of group searching for something to believe in.

I make a plea here for expansion of the mind outwards. To reduce the limitations we have placed on the conscious self. To include other intelligent species on the planet. To at least entertain the idea that they may include the cetaceans—the whales and dolphins.

If this belief is possible for you, perhaps a certain humility becomes appropriate, tempered by the dignity or joy of being part of something much bigger. A part of universal mind—prana—or God.

If you put God outside and set him vis-a-vis his creation in his image, you will logically and naturally see yourself as outside and against the things around you. As you assign all mind to yourself, you will see the world around you as mindless and therefore not entitled to moral or ethical consideration.

The environment will seem to be yours to exploit. Your survival unit will be *you and yours* against them and theirs. Against other races, other species, other religions.

If this is your estimate of your relation to nature *and you have an advanced technology* it seems to me our future looks bleak. We will either die of the toxic by-products of our own hate or simply blow ourselves up.

If that is so, the whole of our thinking about what we are and what other species are has got to be restructured.

If we continue to operate on the premises that have been fashionable in the past and which were especially underlined and strengthened during the Industrial Revolution, which seemed to validate the Darwinian unit of survival, it seems likely we will go in the direction of destruction of life as we know it.

Nobody knows how long we have, under the present system, before some disaster strikes us, more serious than the destruction of any group of nations. The most important task today is perhaps to learn to think in a new way. This new way includes an appreciation of the interdependence of all nature, all species. Human beings are not alone on this planet. The human mind is not the only mind. We as people *suffer* from inter-species deprivation. What a marvelous relief it would be to be able to communicate with another species on the planet. We propose that we are not alone on the planet and we are not necessarily the superior ones on it either. It is time to extend consciousness back into the realms of nature . . . back into unity.

Toni Lilly

PART I

The Ways of Mind

Chapter 1, *Mind, Beginning*, explores the concept of "*virtual consciousness*." Virtual Consciousness is that form of 'knowing' that provides an intrinsic regularity to the basic ways the world seems to work.

In Chapter 2, *Models of Mind*, a brief excursion is made into brain-mind function. Primary emphasis is given to the interpretive consequences of studying the brain from bottom to top, from front to back, and from side to side.

Chapter 3 introduces *The Holonomic Mind* by weaving together findings from the neurosciences, physics, and macrodynamics. The intended outcome is to create a way of looking at mind function as a structurally synchronous consequence of the way the universe functions.

Chapter 4, *Open Mind*, poses the notion that our traditional view of the brain as an organ has created a rather closed-system perception about mind function. The holonomic model extends the interpretation of mind function so that what has been traditionally called para-normal becomes only a variation of that which the brain normally performs.

Chapter 5, *Holonomy*, summarizes the previous chapters and elaborates on the implications of simultaneity and unity between the brain/mind system and natural/planetary systems.

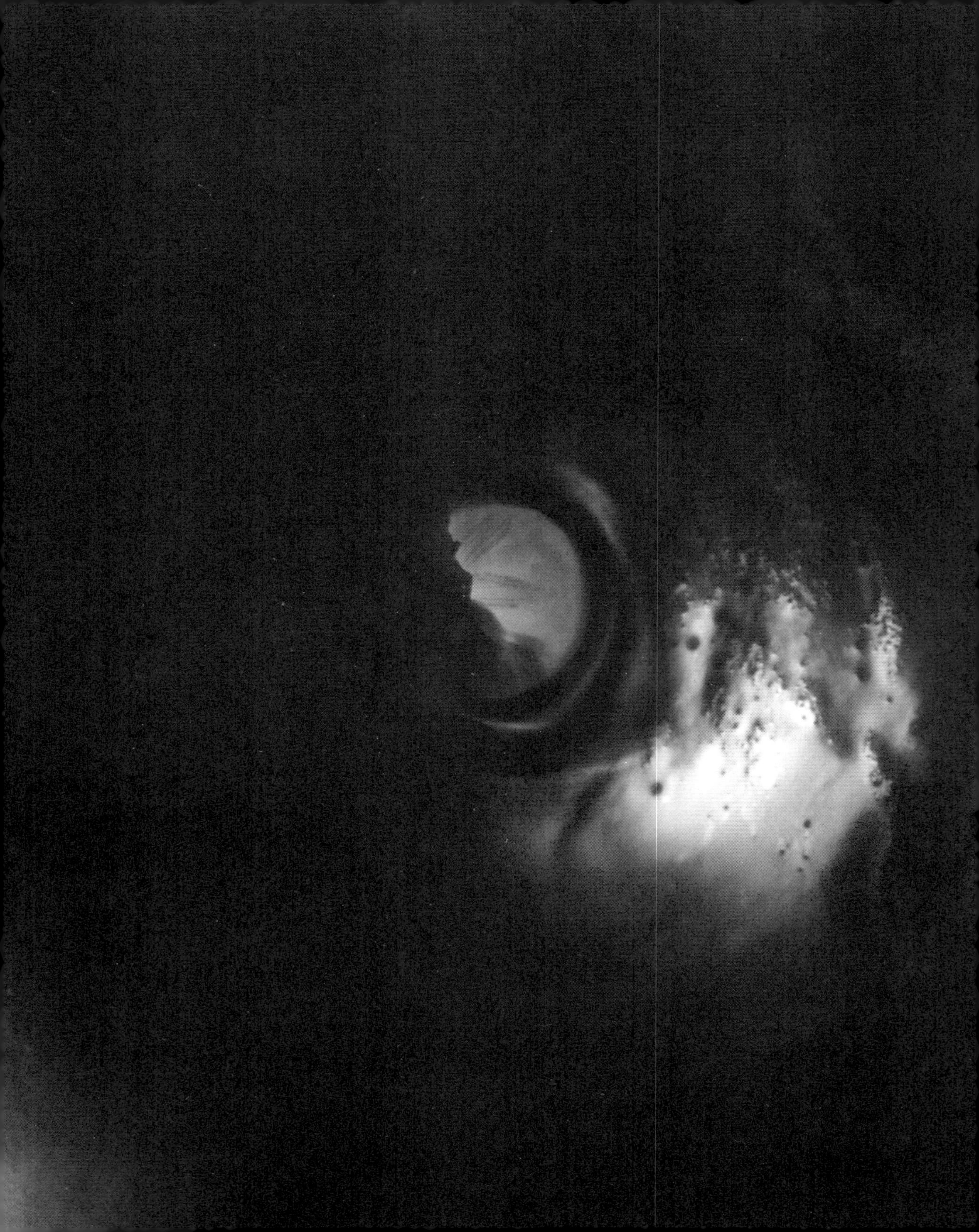

1

MIND, BEGINNING

It was dark, twenty billion years ago. So dark that you and I can scarcely imagine what it could have been like. The universe, stretching to infinity in all directions, was little more than a swirling, drifting mass of the simplest atoms and ions in existence—hydrogen and helium. No rocks, no crystals, not even dust was to be found. Nevertheless, within the universe, there was in this cosmic germ plasm the means to express a form of non-organic life. It had mass. Each particle within this plasm possessed enough substance so that its very presence began to influence other tiny pieces close by. In this non-organic kind of "life," matter is compelled to invite itself into specific kinds of movement . . . into a sort of gravitational dance.

Material stuff does that. By virtue of its possessing mass, it affects other objects. Throughout the universe, substance attracts substance. It doesn't matter whether the substances are gases, solids, or liquids; matter hungrily attracts all other matter . . . drawing its kindred substance toward itself. Naturally, the largest of these cosmic motes are the most effective in drawing smaller bits closer each time they pass. Eventually, some are captured.

The loose material of the universe persistently drifted together, harvested by the silent inward breath of gravity. The inward breath was gentle at first. It was aided by the natural but ill-defined turbulence of the shifting swirl of atoms. Drawn within this inward spiral, certain areas were favored by accretion, and larger and larger masses of hydrogen and helium began to accumulate. The whole process became a persistent and irreversible cosmic dance.

In turbulence, there emerged clotting, deep gravitational breathing, inhaling, exhaling, moving, coalescing, growing.

Was mind born at this time? Was there a kind of consciousness guiding this whole process? "No, of course not," is the indignant claim of most modern humans.

Consciousness is a process of reflection. It is a way of knowing about knowing. The *New Webster Encyclopedic Dictionary* defines consciousness as "the faculty of knowing what affects or what goes on in one's mind."[1]

Implicit within most such definitions is the notion that without humans there would be no consciousness. Consciousness typically is seen to be a uniquely human trait.

Early on, the portion of our species blessed of the Sumerian-Graeco-Roman traditions made a series of policy decisions about mind. These early humans accepted mental reflection as being central to the concept of consciousness. Most claimed that lesser mammals and all other organisms possessed no consciousness. The inferior kin of humans supposedly acted on reflex and animalistic influences. And most certainly, there was no consciousness in earth, fire, air, and water.

Mind and consciousness came to be defined by a capacity for a kind of mental aggression which was in fact directed against itself. Only when the mind fashioned thoughts that intruded reflectively upon itself was consciousness supposed to exist. In this incestual perspective, consciousness most certainly did not exist within the inanimate matter which birthed our solar system.

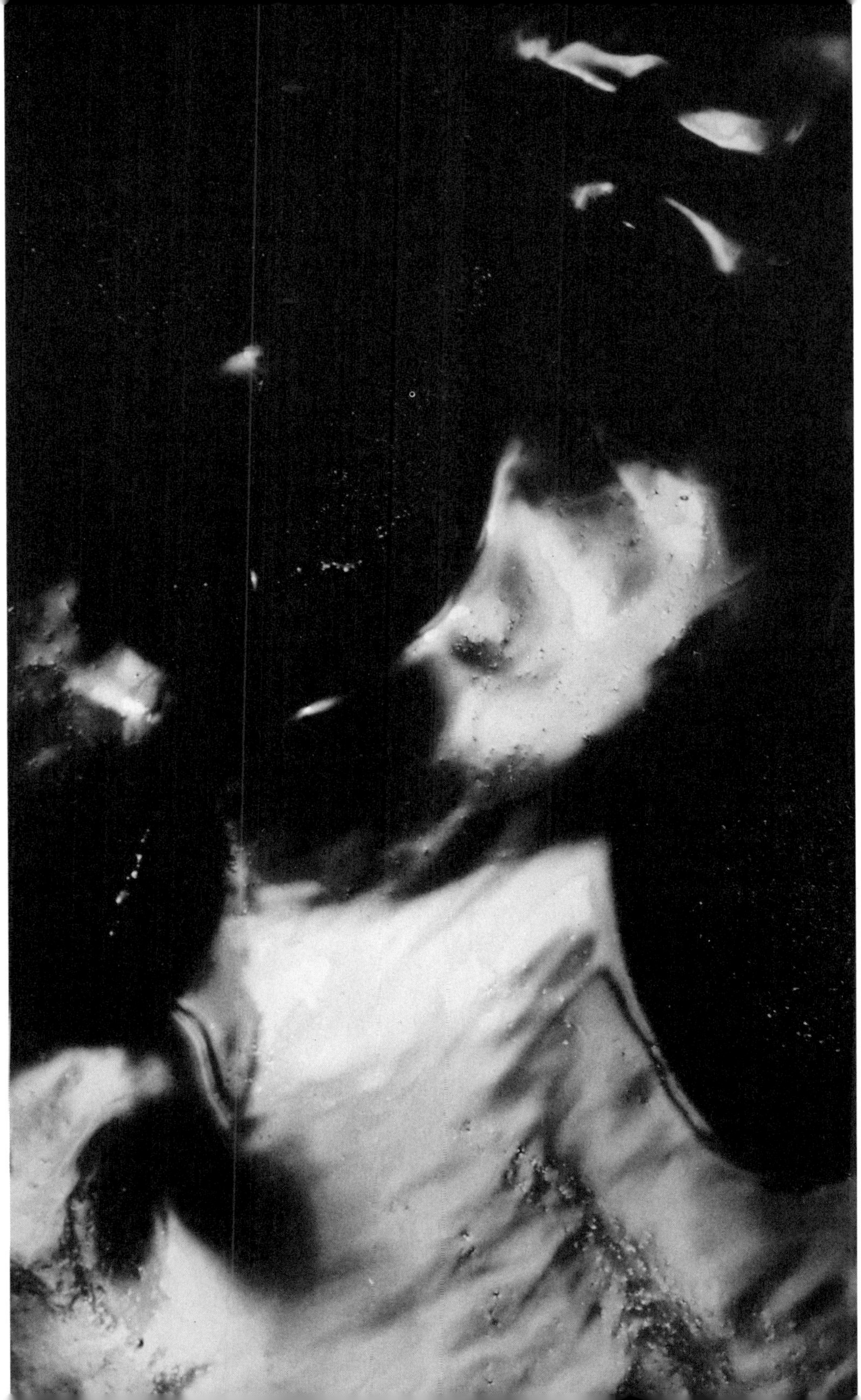

In the infinite domain of galaxies to come, the chaotic swirl began to draw itself into order. The large center of accumulation became larger. In secondary centers far from the main accumulating mass, eddies developed. In these eddies, the material swarmed together to such an extent that long, looping arcs embraced the central region. Eventually, the arcs descended into elliptical orbits. As the granules in these orbital courseways sorted out, they became more stable. Their outward rush was balanced by the gravitational attraction of the huge central mass.

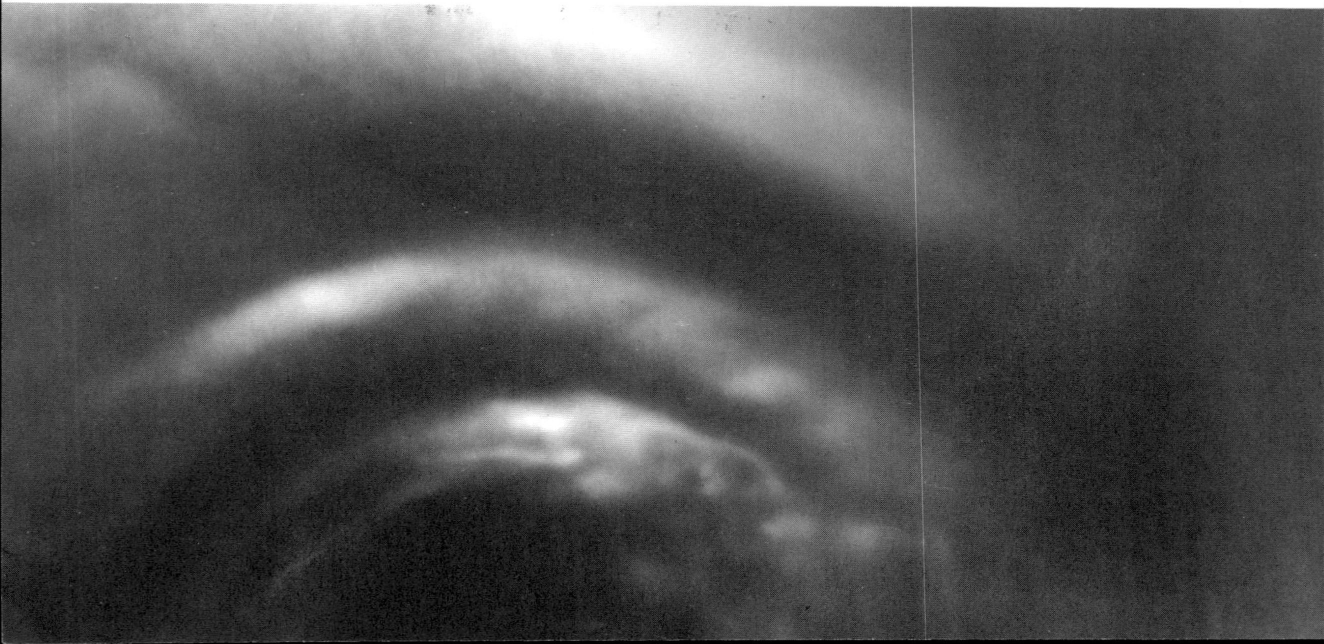

What is gravity? Since the time of Isaac Newton, scientists have been able to predict the behavior of falling objects with great accuracy. Even applied to the domain of space away from the earth, these predictions have been made with fascinating accuracy. Satellites and space probes behave almost exactly as they should, according to the laws of gravity. Material objects attract all other material objects in ways that have nurtured confidence among humans in the domain of human consciousness we have come to call science.

Again, what is gravity? No satisfactory explanation exists. Yet physicists cannot resist an explanation of gravity's presence. To have so accurate a series of *effects* without *cause* creates a discomfort in Western minds that cannot be ignored. To create a context of some comfort, the graviton has been invented. Gravitational interactions are currently thought to be the result of interactions of a subatomic particle which has been dubbed, "graviton." These particles are thought to be massless, chargeless, and, as yet, they have not been detected.

Perhaps the processes of science have given us the basis for a new mythology. The postulation of gravitons—particles which are massless, chargeless, invisible, and as yet not demonstrated—is more comforting than the presence of gods, goddesses, and muses. It is also perhaps more comforting than the postulation of a non-human, non-reflective consciousness . . . non-reflective because there is an inherent, original kind of response pattern which doesn't require thought. Perhaps the human mind—in its centuries-old infatuation with what it can do—has avoided the recognition of a basic kind of consciousness which takes the form of non-reflective consciousness. This non-reflective consciousness appears to establish interaction—whether we recognize it or not—between all matter, and certainly all life forms.

In the swirling but increasingly orderly cosmic disc, the crush of material in the central mass became too severe for it to sustain its original form. The atoms and ions of the inner regions collapsed into a swirling morass of structural oblivion. The inside exploded.

The explosion was muffled. Its outward rush was smothered by uncountable tons of hydrogen and helium pressing inward. Yet the brilliance of the exploding interior's could not be totally contained. Radiation slipped elusively outward. Although most of the energy remained confined in the interior, enough poured outward so that it was clear that our star, the sun, had been born.

In a chaotic rush, the sun's light swept outward, literally blowing less massive stuff from between the reaches of the now several, more or less solid, planets.

All now seemed new. The darkness was swept clean. The new star's energy bathed its children.

Through long and persistent processes, these lifeless shards of space which we call planets developed lives of their own.

If there were some distant cosmic observers, whose viewing style would allow them to watch a million years at a glance, they would see our newly formed planet, earth, as anything but dead. In slow succession, the great ball of accumulated materials began to compress in the center, increase its internal temperatures, and squeeze all forms of solids, gases, and liquids from its interior.

Great welts of volcanoes formed and from them spewed molten rock, water, and a variety of remnant cosmic gases. Mostly, this newly-formed atmosphere was rich in methane and ammonia. Clouds of the mixture formed. Grey, leaden skies and fiery fumes from below mixed with the increased outpourings of radiant light from the sun . . .

The sun. Father Sun . . . enriched the new-born planet with its energy. The sun had now stabilized, and its internal globe of thermonuclear radiance fed from the thick envelope of hydrogen and helium packed in its interior.

The incoming sunlight tore some of the third planet's atmospheric molecules apart, and excited others into blending together. Inside the envelope of methane and ammonia, great storms drifted across the planet earth's surface. Electrical static built from this turbulent caress was relieved when great slashes of lightning racked the scene. In one of billions of such storms, the bursting energy of the lightning and the shimmering light of the sun met molecules which were in just the right configuration for life to form.

And now the planet was truly alive.

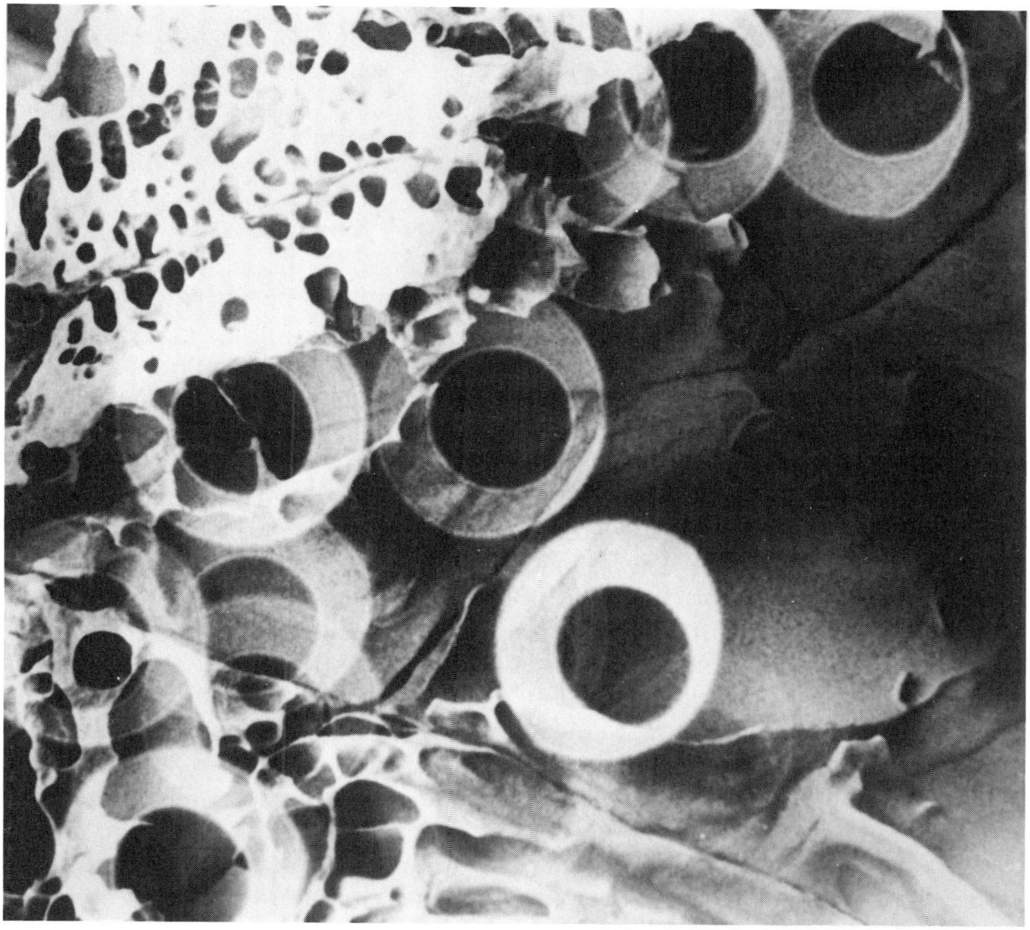

In the murky swamps that were growing in patchy networks across the planet's surface, the first simple molecules that could reproduce themselves were born. They must have "fed" off of other, less clever molecules that had not been laced into life by the sunlight or lightning. Scavenging the aqueous films, these molecules snatched up appropriate configurations and then must have shed key bits and pieces of themselves. These partial patterns followed the imprinted techniques of their parents, and repeated the ways of their own growth. Life was born, and grew, with capacities for renewal.

If reflection and internally-directed thought is necessary for consciousness, then these newly-formed molecules did not fit the conditions. Yet surely they *were* self-replicating. They made themselves over in their own image.

Much has been made of these processes, whether we are discussing these ancient molecule changes, or recombinant DNA. At a fundamental level, molecules make decisions that affect the life that follows. Slight changes in the shape of the template molecules, or subtle substitutions, provide the substance for evolution. Now it is called "genetic drift" and tends to produce slight changes in complex forms of life. But in these primordial seas, profound changes in the organism resulted from these slight variations.

Within the modern cells of humans and most other life forms lives an organelle called the mitochondria. Lewis Thomas, in his amazing and insightful book, *The Lives of a Cell*, speculates that perhaps all complex organisms are a construction of the mitochrondria for the purpose of creating a convenient habitat.[2] Thomas—with a seriousness in his humor—cites the independent mitochondria as being like separate organisms with minds of their own. Even if we accept this as an exercise in metaphor on the part of a serious scientist, the issue of consciousness drifts again across our story line. Is there a molecular consciousness that results in favored choices between atoms and ions? If so, does this consciousness complement the unusual persistence of objects to attract each other gravitationally?

Meanwhile, the hypothetical cosmic viewers which we proposed earlier would have continued to witness the planet's changes. Endless readjustments within the earth resulted in great chains of mountains emerging that were different from the festering volcanoes. These new mountains rose in thick, chain-like welts and helped separate the landscape into great expanses of low areas between the highlands. These low areas increasingly filled with water. The planet seemed destined to separate into land and sea.

These separations allowed all the newly-formed living molecules to gain a kind of isolation from each other. In widely scattered localities, new lightning, new chemistry, and differing resources allowed minute differences to appear. Each life form experienced different opportunities. Remarkable changes were guided at the molecular level. Charges, tiny fibers of electricity expressed as valences, set the odds on what could happen. Life persisted if the ions and molecules that fit the life form's needs were met.

For the first few billion years, the life forms continued to explore the possibilities at the level of molecular chains and rather uncomplicated organisms. But then, new tendencies began to appear. Many of the creatures developed capacities for increasing their odds on surviving. Some began to acquire specialized atoms and ions for the purpose of creating internal and external support structures. Some built needle-like skeletons or plate-like outer shells. In a sense, the first technology was born.

It would be nearly a billion more years of persistent change—sometimes subtle, sometimes dramatic—before the hominids would appear. With the hominids would come the first use of tools, and thus—many would argue—the first technology. Yet was the opportunistic acquisition of a chain of silica molecules to make a spicule, an internal support device, different from the shaping of a piece of silica-rich quartz into a scraper or spearhead?

At one level the comparison would seem ludicrous. "After all," the cultural spokesperson would argue, "the human has a brain; the primordial sponge did not." Who could defend the sponge's intent, its wisdom, or its intelligence? Is the sponge's acquisition an accident of chance, or an expression of consciousness?

Erich Jantsch forcefully postulates non-reflective consciousness in his book, *Evolution and Consciousness.*[3] He calls this non-reflective level of consciousness *virtual consciousness*, and ascribes it to physical systems such as atoms and ions. It is clear that at the basic physical level, matter does show "preferences" in its tendencies to exhibit certain valences, ionic potentials, and "families" of behavior. In fact, the study of atomic physics and chemistry is based on the predictability and expression of atoms' and ions' conformity to certain ways of affiliating. The periodic chart of the elements and the entire domain of chemistry is dependent upon such electromagnetic "preferences." To ascribe Jantsch's label of virtual consciousness to such tendencies is then only a matter of taste.

In a sense, this question of taste has to do with world view. Our traditional, Western world view tends to operate in reductive modes. That is, we are prone to breaking larger perceptions down into component parts. We have an affection for pieces, rather than wholes. The outcome is generally a kind of philosophical chauvinism that ends in an elitism of reason. Consider this statement, *Creatures that reason have consciousness*. By our definition, this separates us from all other creatures and the material stuff of the universe. The outcome of this is that we establish domain over consciousness, *we put ourselves in a position to exploit all other life and matter.*

Remember that Jantsch's virtual consciousness is a non-reflective consciousness. This description in itself may be chauvinistic, as we are not certain whether atoms and ions reason reflectively or not. But, absolving ourselves of that argument, it is useful to accept the persistence of electron courtship as a reality. Atoms and ions *do* behave in fairly predictable ways . . . predictable enough to be expressed as profit in all the major corporations on this planet. Clearly, electromagnetism—like gravity—is consciousness . . . at least the kind of virtual, non-reflective consciousness postulated by Jantsch.

Thus, the cosmic drama of planetary conception, gestation, and birthing—all guided by the presence of gravitational force, and textured by the interconnecting preferences of specific atoms and ions—becomes a choreography of consciousness.

Along with these two—electromagnetism and gravity—other forms of consciousness accompany the basic stuff of the cosmos. These are termed the weak forces and the nuclear binding forces, or the weak and strong nuclear forces. Classically thought to be made up of like-charged particles, the centers of atoms—nuclei—were puzzling. While negatively-charged electrons swirl in changing but orderly clouds around the nucleus, the nucleus itself is made up of clutches of like-charged particles. Their likeness of charge should result in an internal repulsion that would shred them to bits. But most nuclei *are* stable. Fritjof Capra, in his book *The Tao of Physics*, says:

> "Nuclear matter is thus a form of matter entirely different from anything we experience 'up here' in our macroscopic environment. We can perhaps picture it best as tiny drops of an extremely dense liquid which is boiling and bubbling most fiercely."[4]

Is it not a kind of consciousness that tends the nuclei and allows these stable centers of atoms to marshal the electromagnetic clouds of electrons into service? The electrons which surround the nucleus in turn create webbed networks of globular structures with other atoms and ions. In this way, molecules are born. Once formed, the atoms—but more likely, the molecules—begin to pulse to the presence of other molecules, until more elaborate forms emerge. As mass increases, the subtle presence of gravity extends its invitation to dance.

Woven together, these forms of consciousness are the basis for this book's concept of planetary consciousness. Electromagnetic, strong and weak forces, and gravitational forces are the first expressions of the virtual consciousness of our shared unity with the planet. Their presence, their persistence, and their processes are the sculptors of all that is.

Each human—Australopithecus, Genghis Khan, Marie Curie, Cochise, Martin Luther King, Margaret Mead, and the writer and reader of these words—owes their being to these forms of consciousness.

Planetary consciousness is a form of spiritual unity that may re-extend our consciousness to the inanimate, as well as the organic, on our planet. Planetary consciousness requires a re-establishment of the honoring of the mother . . . the true spiritual mother of all of us . . . the planet.

Planetary consciousness then must be expressed by more than our willingness to pay attention to recycling and pollution control. It must be expressed by more than membership in conservation organizations and macrobiotic diets. A return to planetary consciousness clearly may be marked by efforts in all the ecological commitments cited above. But even more central will be a reinstatement of the forms of consciousness that return us to our origins. These origins will invite us again to know and honor virtual consciousness. We will know again the ways of knowing of the sperm and the egg, the embryo in its amnionic ocean. The way our evolutionary journey to this very day is a testimony to all the life the planet has known. Beyond that . . . such a return to planetary consciousness will invite us into a oneness with the journey of the planet itself.

Perhaps we may even know just as organic life on the surface of the earth knows; we are part of its story and it is part of ours. We may join again in the world view of the cultures before us, and those among us today, whose technologies have not ravaged the planet. We may extend our capacity to honor all the forms of consciousness as expressions of a more holistic vision. Such holism is an invitation to include ourselves in the cosmic vision . . . *include* ourselves rather than exclude ourselves, as has been our impoverished pattern through history.

For centuries, Western philosophy and Western spirituality have excised humans from the planet. The earth—its wilderness, its resources, its life—has been provided for the purpose of exploitation. Its surface to be leveled, paved, mined, and changed; its life to be slaughtered, or tamed and coaxed into more "productivity"; its air and water to be marshalled into further growth and energy production. The legacy of this private ownership of consciousness has been awesomely expensive.

Consciousness is not the exclusive domain of *homo sapiens*. It is intrinsic to all things. Consciousness is part of the planetary presence, and its restoration for us may well be the first necessary step toward a mutualistic survival. Mutualistic in that it must honor all elements of what has become an injured womb . . . an injured planet.

Planetary consciousness may allow our varied forms of consciousness to coalesce. To unite. Our minds may reflect these ways of knowing. We may honor again and be nourished by the mind of our mother.

2

MODELS OF MIND

Brain research provides comfort for the Western mind. It provides us with "objective" data about some things which intuitively we already know. With measurements in the form of weights, volumes, electrical productivity, chemical balance, and response to electrical and chemical stimulation, a mountain of data has emerged in the neurosciences. Though singly and collectively these data tantalize as much as they fulfill, they *do* provide a greater sense of wisdom with respect to brain function. And yet—even armed with all they yield—most neuroscientists are left unfulfilled.

Very little has brought resolution in the millenia that have passed since humans first recorded their interests in this area. Hundreds of thousands of pages have been written about the mind-brain dichotomy, and as much about the brain-mind-consciousness issue. Philosophers, psychologists, surgeons, clerics, and mystics have entered the fray. Most have attempted to trace a logical course to create an argument that makes "sense" within the reductive framework of Western thought. Yet all of these efforts come to little agreement.

The philosophers smile and cite the *Heisenberg uncertainty principle*. This principle is central to the concept of uncertainty in measurements made in the world of subatomic particles. What it means is that one cannot measure both the velocity and the position of an electron. The impossibility of the task is linked to the notion that the energy required in making the measurements for one condition—the velocity—affects the electron's position, and vice versa. What this really means is that the observer/measurer is not an objective entity at all. In reality, the observer/measurer is a participant in the enterprise.

So it is with the brain-mind-consciousness issue. Using one's brain to create one's mind, and then reflectively think about itself to create consciousness, ensures a kind of interrelated action that may never be unsnarled scientifically or philosophically. Suffice it to say that one's biases about the structure, function, and evolution of the brain affect each of the aspects of interrelatedness.

Until recently, the most popular way that the brain was studied was from bottom to top. Early on, neuroscientists concluded that the central nervous system—as extended via the spinal cord and its dendritic nerves—was the basic, rudimentary brain in humans. Such nerve systems represented all that was to be had in nearly all of the simplest bilateral life forms. When organisms of higher complexity appeared, they possessed a flexible rod of cartilage along their backs. This was called a notochord.

Clearly the bottom-to-top neuroscientists considered this root-like appendage dangling into the body from the base of the brain to be secondary to the neural magnificence found in the skull. The hindbrain and midbrain were also relegated to more "primitive" status. It was on the forebrain and its specialized enlargement, the cortex, that they focused their attention. In essence, the sequence represented the transformations in complexity from annelid worms through fish, amphibians, lizards, birds, and lesser mammals. *Homo sapiens* has the largest cortex of all these creatures. And thus, as perhaps a non-intended form of chauvinism, the cerebral cortex emerged as the primary criterion for significance in brain structure.

That we still possess all the vestiges of this evolutionary journey seems to have escaped notice as being significant by most neuroscientists. The message of evolutionary grace seems too direct. The huge cortex and enlarged midbrain proved so seductive for human emphasis that we tended to forget the forms of evolutionary consciousness that preceded the development of the cortex's reflective consciousness.

Steven Rose says in his book *The Conscious Brain*:

> Behavioral determinists have maintained a 'fish brain' and a 'frog brain' which are in many ways more important . . . than the much vaunted cerebral cortex . . . But to extrapolate from these facts towards the claim that because similar structures exist in men and frogs, man's behavior is inevitably frog-like, is nonsense.[1]

Rose's posture is in clear opposition to many behavioral determinists such as Lorenz, Tinbergen, and Ardrey. These ethologists are quick to conclude that earlier forms of consciousness dominate cortical functions. Both views may overstate the loss of a facet of planetary consciousness. The behavioral determinists want to interpret our behavior through that of fish and frogs, while the cerebral celebrators would argue we have transcended such behavior. In my view, both postures divert us from a vital reality. It is neither instinctive behavioral control nor a form of transcendence erased by evolution that relates to consciousness here. Instead, *it is the reality that these kinds of structures persist in our brains, and that we have access to modes of consciousness embodied in these structures.*

So, as we acknowledge the emphasis of the vertical or bottom-to-top students of the brain, I feel we have a model that clearly invites us to widen our vistas of consciousness. We have within us the capability of returning to the wisdom of those more closely allied to the planet. We have within us the capability of returning to the wisdom of those whose perception of reality is more closely linked to the consciousness of the planet. Most ethologists tend to denigrate this consciousness by emphasizing the primitive and animalistic behaviors of preceding species. The spokespersons for *sapiens* bias us with ego. Reason, logic, and consciousness—they claim—separate us forever from our baser ways of knowing. But is it the ways of knowing, the consciousness, they want to separate us from . . . or is it the behaviors?

The cerebral cortex emphasizes in humans the capability for processing relationships. Much of this processing is related to the association areas of the cortex. Figure 1 shows the emergence of the association cortex in a diagram that compares a human brain with the brains of a rat, a cat, and a monkey.

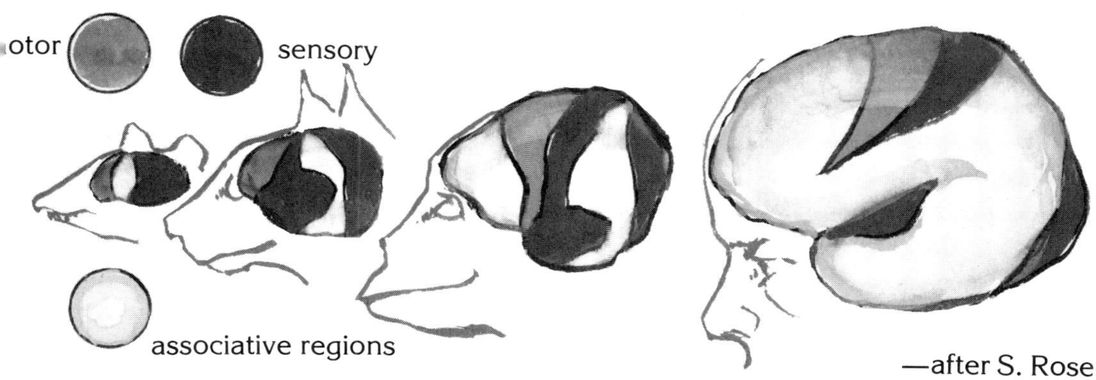

Figure 1.

—after S. Rose

Figure 1 illustrates clearly the remarkable increase in the associative regions of the brain, those areas most favoring reflective thought. Researchers point out that in humans not only is the size of the associative area relatively and quantitatively greater, but the higher convolutions of the surface of the brain show a remarkable increase in the total number of neurons or brain cells as well.

The increase in the associative areas leads one to leap to the conclusion that there has been a comparable decrease in the motor and sensory areas—the areas that best suit the consciousness styles of monkeys, cats, and rats. Yet if we look beyond the proportional relationships, it looms as a reality that humans have more brain volume and more *total neurons* in the service of motor and sensory capacity than do monkeys, cats, and rats.

Perhaps then, our lesser ability in the motor and sensory domains is as much a function of what we consciously pay attention to as it is the result of lost capabilities. In other words, we—as members of a Western technocratic culture—have *chosen to use specific brain capabilities in ways that interfere with other, less attended, capabilities.*

All one needs to do is compare two humans whose cerebral capabilities are equal in design and yet are of awesomely different consciousness. A stockbroker from San Francisco and a hunter on the Kalahari would provide a perfect pairing. The stockbroker would be guarded and restrained in normal body expression, while the Kalahari hunter would cherish a more diverse repertoire of appropriate movements. One would move in halting, restrained fashion; the other in fluid, open ways including unquestioning use of crawling, creeping, and full-out running. The broker would cultivate senses more isolated from nature, while the hunter would exercise senses developed through immersion in natural systems.

This leads to a notion I first heard offered by Karl Pribram: consciousness is what we pay attention to. In my interpretation, the possession of an elaborate cerebral cortex does not deny *homo sapiens* the stopover places in the journey from the origins of the planet to the development of human beings. We need not allow ourselves to be imprisoned by specific cortical capabilities at the expense of others. Perhaps technological humans have created an ecology of consciousness that denies full use of the brain. But the brains of technological humans are not so constructed as to isolate us from our full potential. It is a matter of choice.

The cortical choices we have are nowhere better exemplified than in another model of neural function—the lateralized model—one which emphasizes side-to-side specialization in brain function. Less than fifty years ago, a medical student would have had to search diligently for any information about the lateralization of function of the cerebral cortex. Ten years ago, it would have been easier; perhaps a hundred papers and a few chapters in books would have been available. Now it requires nearly a full-time perusal of literature to keep up with the annual production of research findings.

The pioneering work of Roger Sperry, Joseph Bogen, and Michael Gazzaniga has led to an explosion of interest that has strongly influenced fields as diverse as education, corporate management, advertising, and psychotherapy. Robert Ornstein's 1972 book, *The Psychology of Consciousness*,[2] has been the single most influential source in the popularization of this model. Following its publication, the phrases "right brain" and "left brain" became new metaphors for many people interested in mind function. My own book, *The Metaphoric Mind: A Celebration of Creative Consciousness*,[3] emphasized the unity of mind by attempting to bring the functions of the "right brain" into respectability, while not disowning nor dishonoring the "left brain."

My first interest in the area came in 1961 when I worked with Jerome Bruner, then a Harvard psychologist who had just published a book entitled, *On Knowing: Essays for the Left Hand.*[4] Bruner's metaphors of "right-handed" and "left-handed" knowing turned out to be precursors of the right- and left-brain metaphors. It is important to introduce a little vertebrate quirk here. In higher order vertebrate mammals, there is a cross-over design involving the nervous system and the brain. The left side of the cerebral cortex controls—or at least powerfully influences—the right side of the body. The right cortex governs the left side functions. Thus, Bruner's *Essays for the Left Hand* was in fact a book about what some would call "right-brain knowing."

As research continues, information is growing in volume and complexity. Much of the research related to lateralization is now conducted with the aid of biofeedback machines. These monitors are able to determine what kind of energy is being generated by the brain. With electrodes attached to both sides of the head, one of the more popular biofeedback devices indicates which side of the brain is most "active" when certain tasks are being performed by the person being tested. Beta (β) energy production is thought to be the most aggressively active mode. Alpha (α) production is considered a more relaxed, less active state. In most adults in technological cultures, the left cortex appears to be specialized to engage actively when consciousness is directed toward logical, rational, linear, sequential, and time-ordered modes of knowing. The right hemisphere becomes engaged more actively when analogic, metaphoric, intuitive, holistic, and visual/spatial tasks are performed.

Because the cortex is the dominant arena of the associative functions, let us consider for a moment what lateralization means. If a person were to experience a flower—a lilac, for example—then the lateralized brain would begin to process the experience. The formal, more logical, associations would likely be initiated and sustained by the left cortex. The patterns that interrelated "lilac" with formal experience would begin. Perhaps the scientific name of the flower, its family relationships, the number of petals per blossom, and its leaf configurations would filter into consciousness. The right cortex might wander across thoughts linked to smell, springtime, family gardens, deserted farmsteads in the Rocky Mountain west, and a kiss that followed the handing of a lavendar bouquet to a lover.

Debate exists as to whether this specialization is neurophysical, attitudinal, or a combination of both. There is also much harrumphing and evasive qualification by neuroscientists who feel that these findings are being over-simplified.

"Dichotomania" is one term coined by those who feel that the popularization of these ideas is resulting in a glib, undisciplined application. And rightfully so. We all live in a culture so enthralled with its technocratic rampage that we are imprinted from birth with dichotomistic thought. In a strange and tacit way, we are taught to take the digital tact and to reduce life's options to on-off, up-down, right-wrong, republican-democrat, and a host of other dyadic contexts. For many, the search for clarity nurtures the two-choice perception.

None of the serious students of lateralization encourages such on-off extremism, contrasted by the more exploitative style of some of the popularizers. It is clear, for instance, that for a quality of knowing to be called left-brained, it requires beta (β) wave neuroenergy production in the left cortex. Some assume that when the left brain goes "on", the right brain goes "off." Nothing is farther from the truth. The right brain may go into alpha (α) wave generation. Alpha is an awesomely productive mind state. It is characterized by receptivity and acceptance. Thus, it is likely that such a mode provides a means for the right hemisphere to experience tacitly that which the left hemisphere, in beta, is processing. Whereas the left mode is actively inclined to be reductive, the right tends to be proliferative. While the left funnels options toward *a* solution, the right might be preparing for the exploration of dozens of solutions.

It is likely that right-brain knowing—when actively expressed by beta activity—is offering patterns of legitimate options, as opposed to focusing on a single option. In non-discriminatory ways, the mind's attention flows across mindscapes of unorganized and loosely-structured wisdom. Essentially, the mind doesn't care where it goes or how it gets there. Meanwhile, the left is in alpha. In a sense, it is preparing to make sense of the options created by the right.

Michael Gazzaniga, one of the original neural researchers in this field, was so struck by the tendencies to turn these qualities into cliché that he wrote a somewhat impassioned statement for equilibrium. His 1978 book *The Integrated Mind,*[5] cites a remarkable body of evidence to support the concept that both sides of the brain's cortex address all tasks. Gazzaniga relieves us of the content-centered dispute. Many people have thought mathematics and symbolic language to be linked solely to left-cortical ability. Art, music, and dance were claimed by these same people to be the content of the right hemisphere. Gazzaniga and others have shown clearly that is not the case. One cannot confuse processing functions and content. The brain can only lateralize the way it processes experience; it cannot lateralize the content of the experience.

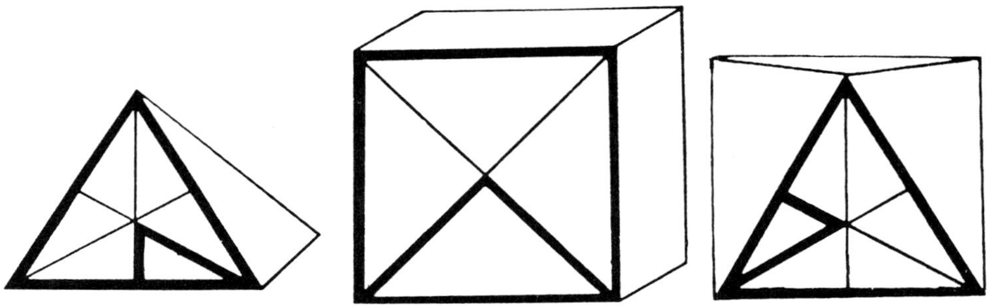

Some mathematicians exhibit strong preferences toward right-brain function when performing their "art." Just as surely, many artists exhibit strong preferences toward left-brain knowing when defending their visual logic. Again, the issue is consciousness. For me, the lateralized model is not so much a case of conjoint storage bins of our forms of knowledge as it is an expression of neural efficiency. Lateralization is an expression of cortical efficiency.

Cortical effectiveness is another issue. Effectiveness is linked to consciousness. To return to the statement made earlier—that consciousness is what we pay attention to—it becomes clear that at the attitudinal level, what we pay attention to is an ecological issue. What this means is that the natural/cultural ecology that humans experience is most certainly effective in determining what humans pay attention to. Let's reconsider our San Francisco stockbroker and Kalahari hunter. Each surely scans different cerebral qualities. The broker's culture has influenced the programming of the associative areas of the brain to favor media of abstraction; that is, words, letters, numbers, and other symbols. The hunter is far more disposed to associative areas wedded to motor and sensory expertise. One "expert" reads the written codes of the finance pages; the other "reads" the unwritten language of spoor.

Western technological ecologies favor left-brain consciousness. Natural technological ecologies like those of the Kalahari hunters favor right-brain consciousness. Thus, left and right brainedness—as tarnished as this metaphor may be—is, to a significant degree, a matter of choice. We *choose* what we pay attention to.

It is clear that evolution may inadvertently have created the chauvinism that resulted in our exaggeration of the cerebral cortex and its attendant associative functions. Simultaneously, it is clear that the cultural ecology within which we live may bias us toward the sides of the lateralized cortex on which we focus our attention. Yet, even though these tendencies exist there is nothing to prevent our re-extending our consciousness freely to both sides of the cortex and to the brain's regions below the cortex to counteract the *sapiens* chauvinism. In simpler language we can—at will—choose to extend our consciousness by paying attention to and honoring all the modes of knowing preferred by the cortical brain, and all the rest of our brain-mind system as well.

The evolutionary process has done nothing to excise planetary consciousness from us. The acculturation process has.

We have little choice about either as history. Natural history has given us the body-brain system that we have. Cultural history has given us the mind-consciousness system we have. Our first gestation is in planetary consciousness; our second is in cultural consciousness. For those of us immersed in Western, technocratic emphases, the models provided us by culture have systematically excised us from planetary consciousness. There exists the tacit assumption that we have evolved *away* from earlier modes of consciousness. This is not the case. *We have not ascended; we have co-existed.*

Ernst Cassirer states it this way: [Darwin] "destroyed the arbitrary limits between the different forms of organic life. There are no separate species; there is just one continuous, uninterrupted stream of life."[6] Life, not form, is that which establishes unity. Life is what binds the minds of species into a single context. Our cortical capabilities give us more options for compassion, and yet we express the tendency to reduce compassion by the pretense of supremacy.

The lateralized model assures us of the possibility of a kind of salvation. Strangely without content or form, the left-brained and right-brained ways of knowing may well persist to retain planetary consciousness. The left brain—parented by the logic of culture—is complemented by the right brain, with its preference for holism. The left brain invents closed systems. The right brain retains openness. In culture there are closed systems, but in nature there are none.

Must we rest then with deficit models? The commission of vertical brain study produces the chauvinistic mind, and the lateralized model coaxes us toward dichotomy. Both approaches seem to invite us attitudinally, and at the level of consciousness, to take sides, to choose as a child of adversity.

If we absolve ourselves of the chauvinism inherent in the vertical model and trend toward a conscious equilibrium in the lateralized model, then another option appears. One filled with simultaneity, holism, and non-discrimination in the arena of consciousness.

More than this, the new model is natural as well as cultural. It is imbued with the notion that planetary consciousness—or for that matter cosmic consciousness—is fundamentally human. This model, the *holonomic* model, gives humankind one of the first contexts of unity with all that is. In holonomic knowing, we re-enter the premises of inclusion with all that exists.

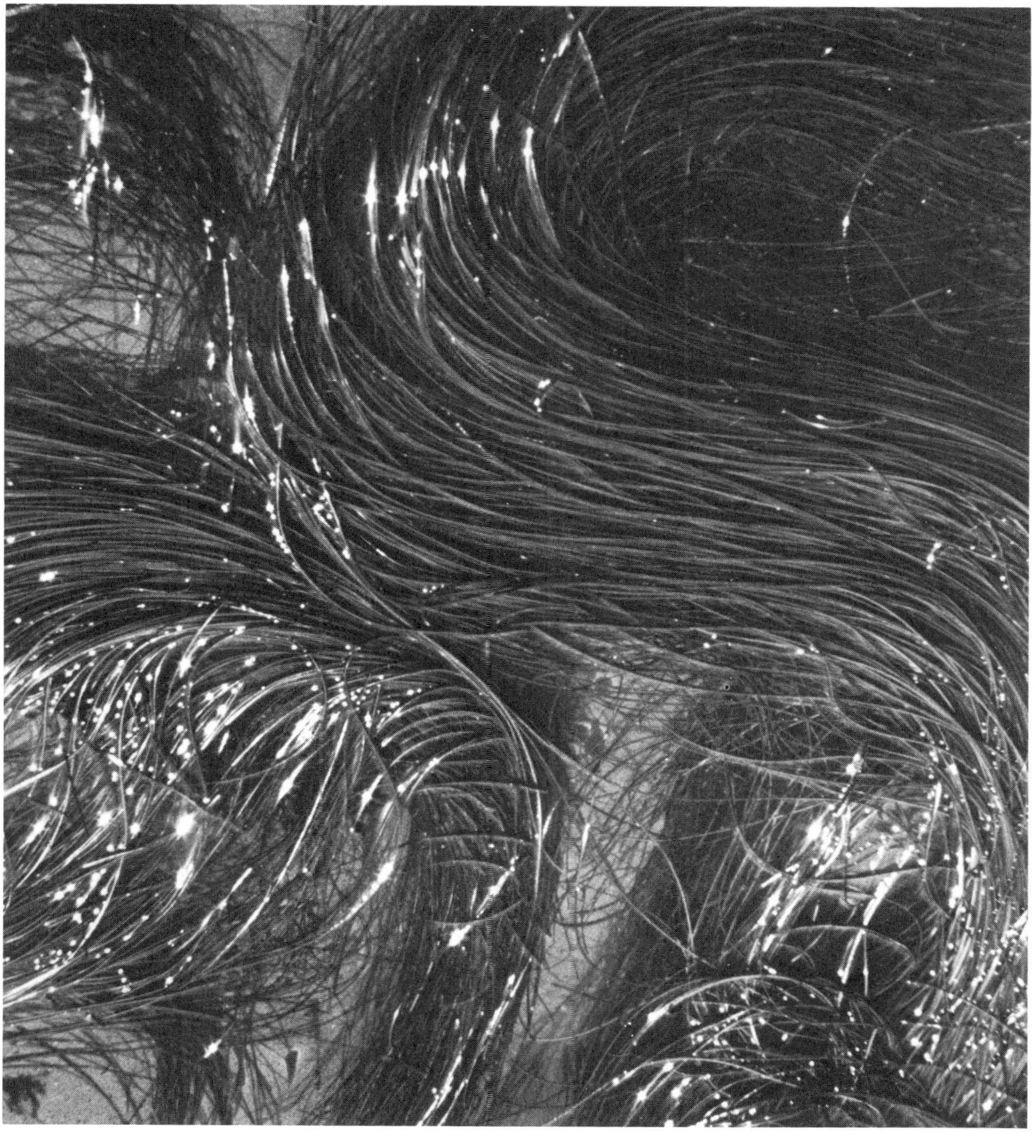

3

THE HOLONOMIC MIND

Holonomy is born of two parents: *holos*, meaning whole or entire; and *nomy*, meaning science or systematic study. Holonomy then becomes the science or, as I prefer, the systematic study of wholeness.

George Leonard originally defined holonomy in the popular press in his book, *The Silent Pulse*. In his view, "holonomic means named for or in the nature of a hologram."[1] For me, this definition puts the child before the parent. The hologram is simply a special case of the concept of holonomy. I believe it is important to realize that the *hologram* is a physical metaphor that expresses the concept of holonomy and holonomic contexts. However, it is also important to remember that the hologram is but a shard of a far larger modality of being. I will detail the role of the hologram as a metaphor later in this chapter.

Holonomy as a practice in Western thought died an early death on the altars of mind in the Sumerian and Graeco-Roman traditions. The philosophical treatises of the Greeks in particular stressed reductive analysis. The Greeks described both deduction and induction. Deduction is reasoning from holism to the specific. Induction moves from the specific toward the holistic. Although the purpose of both was to augment the other, it appears that the concept of deductive prowess influenced the image of the keen mind, and in turn became the Holy Grail of Western thought. A person who functions effectively in the deductive mode is considered intelligent. As with Sherlock Holmes, each step prepares the way for the statement, "Elementary, my dear Watson." Centuries passed in which the dominant products of thought were classification schemes, taxonomies, and overwhelming arrays of pigeonholes in which the earth, its life, and all the qualities of humanness were tidily inserted. Voices like Chaucer, Shakespeare, and Blake were among those who created testaments to a loss of wholeness.

Several attempts to resurrect holonomy failed. Yet there was so much reductivism ready to emerge . . . too many pigeon holes to fill. The Scientific and Industrial Revolutions merged to extend the exploitative technologies of the Agricultural Revolution. Humans were providing themselves with technologies that would separate them from the earth. The tools of mind and substance created by science and industry shifted human lifestyles from the desert, forest, and plain's ecologies to the urban. A result was a shift in human modes of thinking rich in open-system, natural metaphors to ecologies that nurtured the closed-system metaphors of urban culture. Human beings were moving from nature and its open systems to cultural settings abundant in the closed-system premises of law, science, religion, and formalized relationships.

The issue of open- and closed-system thought is important here. While it is true that no system can be created completely closed, it is clear that some systems are more closed than others. For example, consider the difference between the eating habits of a bird—an open-system design—and the consumptive appetite of an internal combustion engine, a closed-system design. Let us say the bird is a seed eater. It has the choices of nearly a million kinds of seeds in the path of its home territories and migration routes. The internal combustion engine is afforded few choices; its design, a product of closed-system thought, precludes all but a few options.

The open-system design has infinitely more options than the closed-system engine. In addition, the open-system design is independent and able to respond to random encounters. Also, for short periods of time, seed-eaters can eat non-seed food such as suet, insects, and even carrion. Open systems operate on the premise of inclusion, whereas closed systems operate on the premise of exclusion. Moreover, the metaphors of open systems tend to be self-generative, repatternable, and imbued with holism. Closed-system metaphors are digital and linear. Open-system metaphors lead our minds into open-system thought. Closed-system metaphors lead our minds into closed-system thought.

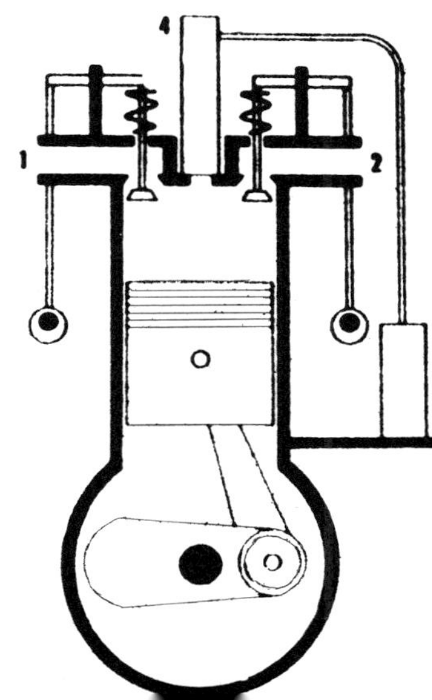

What has this to do with holonomy or a holonomic vision of mind? Holonomic systems are open systems. Both of the models of mind that were discussed in the previous chapter have strong biases toward closed-system interpretation. The vertical model invites closedness with its implications of linear, exclusive evolution. Once the cortex attained human proportions and the associative functions dominated, the principle of exclusion was enacted. All the brain structures below the cortex were demeaned, if not denigrated. Surely we still admit to the midbrain and the lower brain, but these do not have the position of honor that the cortex possesses.

The lateralized model of the brain invites closedness because it is digital. It invites us to think of one half of the brain as on (β) while the other is "off" (α). Notwithstanding the efforts of Gazzaniga and others to emphasize the integration and interdependence of the cortical hemispheres, the lateralized model is still seen as a digital model. Moreover, it emphasizes the functions in a way that nurtures dichotomy; and dichotomy is a closed-system context.

The holonomic model that I propose here is a synthesis of a variety of sources. Most notable for popularizing the substance and vision behind the holonomic model are Karl Pribram and David Bohm. Pribram's extensive experience as a neurosurgeon, neuroscientist, and avid researcher has provided him with a wealth of insight that makes this model both theoretically and practically viable. Bohm, as a physicist, has approached the issue from his studies of consciousness and physical systems. Throughout his explorations—whether in subatomic or cosmic dimensions—Bohm is a champion of unity and wholeness.

Both of these men step boldly beyond the closed-system contexts common to both neuroscientists and physicists. Pribram is both adamant and informed about the holistic capacities of the brain. Bohm sums up his convictions in the Summer/Fall 1979 issue of Revision Quarterly, as he says:

> The true state of affairs of the material world is wholeness. If we are fragmented, we must blame it on ourselves.[2]

Both Pribram and Bohm represent the latest stages of a paradigm shift that began in the early 1900s with statements about the brain by William James. James maintained that the brain acts as a filter to a larger reality. In effect, he defended consciousness as the way we select what we know. Three years later, Einstein published his theories. In 1929, the stockmarket crash focused popular attention in such a way that it all but eliminated our recalling a revolutionary claim by neuroscientist Karl Lashley. Lashley claimed that memory is not relegated to a specific part of the brain. He claimed instead that memory is encoded *throughout* the brain.

In 1947, Dennis Gabor created the mathematical models that eventually helped Emmett Leith and Juris Upatnicks create holograms from laser light sources. Since then these ideas have been primarily the domain of Pribram and Bohm. They have explored—separately and together—models of the brain and the universe based on the hologram.

The hologram is, for both Pribram and Bohm, the metaphor that best represents their conception of the way the brain works. As complex as the concept of holography is, a rather concise statement was made by Marilyn Ferguson in her *Brain/Mind Bulletin*:

> Holography is a method of lensless photography in which the wave field of light scattered by an object is recorded on a plate as an interference pattern. When the photographic record—the hologram—is placed in a coherent light beam like a laser, the original wave pattern is regenerated.[3]

No lens is used in the process, so the resulting three-dimensional images result from wave frequencies uniting in a coherent form. Furthermore, each part of the hologram can create the entire image.

The very notion of lensless photography is generally anti-intuitive to one trained in the Newtonian tradition of lenses and lens systems. Yet even the most cursory reflection can dispel this dis-ease. When we look at objects before us, such as the words on this page, it is clear that good old Newtonian physics suffices to get the word-images from the page through the lenses in our eyes to the "screen" in the back of our eyeballs. But from there lenses don't work. The optically-excited nerve cells turn the image into electrochemical impulses that begin the process of sweeping the images in this transformed shape back across billions of non-optical nerves. Almost like neuroelectrical windstorms, the content, effect, and tacit implications of the one-time images ripple across our consciousness.

Once past our eye, all visual experience is lensless. It is no more than the interaction and interrelatedness between waves, particles, and charges. There are no pictures in our brain. There are just the pulsing winds of our minds exciting a fullness and a richness to the visual impression of a non-visual experience. To understand this, one must simply witness the anxiety of the transparency film maker who switches over to the use of video or electromagnetic tape. There is some comfort in holding up yards of transparency film and seeing real pictures on it. This comfort is not experienced when one looks at video tape. Yet the same images may exist on both.

In the figures that follow, the holonomic model is explored.

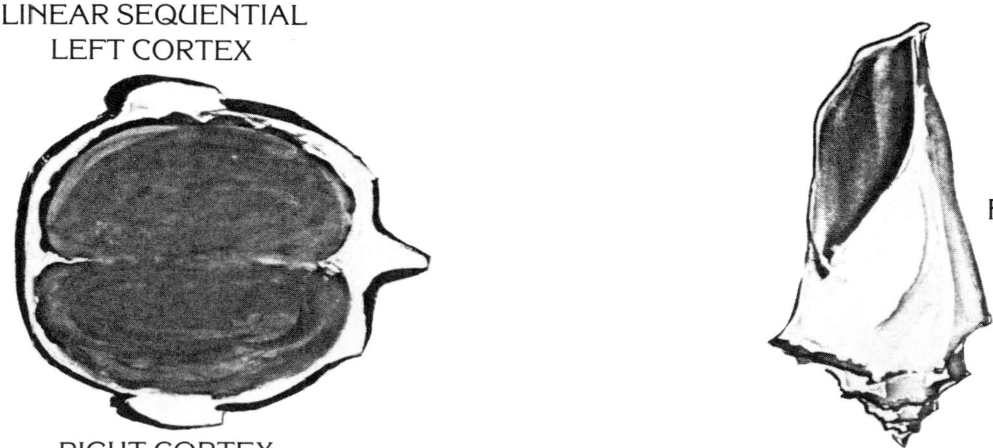

Figure II

The most simplistic version of the lateralized model is shown in Figure II. It is clear that Galin, Gazzaniga, and others would shudder at this interpretation. But inherent in this simplification is the most potent weakness of the lateralized model. As the brain is searched for "shellness," it is clear that the logical, linear, sequential expression of shellness, i.e., conch, is not *all* that the left hemisphere has stored in memory. Just as surely, the right hemisphere activity has not been limited to the visual-spatial configuration of the shell.

Actually, if one could inventory the left hemisphere, one would find all sorts of more analytic associations of the *whole* experience related to the specific shell and shellness. Similarly, the right hemisphere has encoded associations that go to great depth and breadth in regard to shell and shellness. Figure III illustrates this.

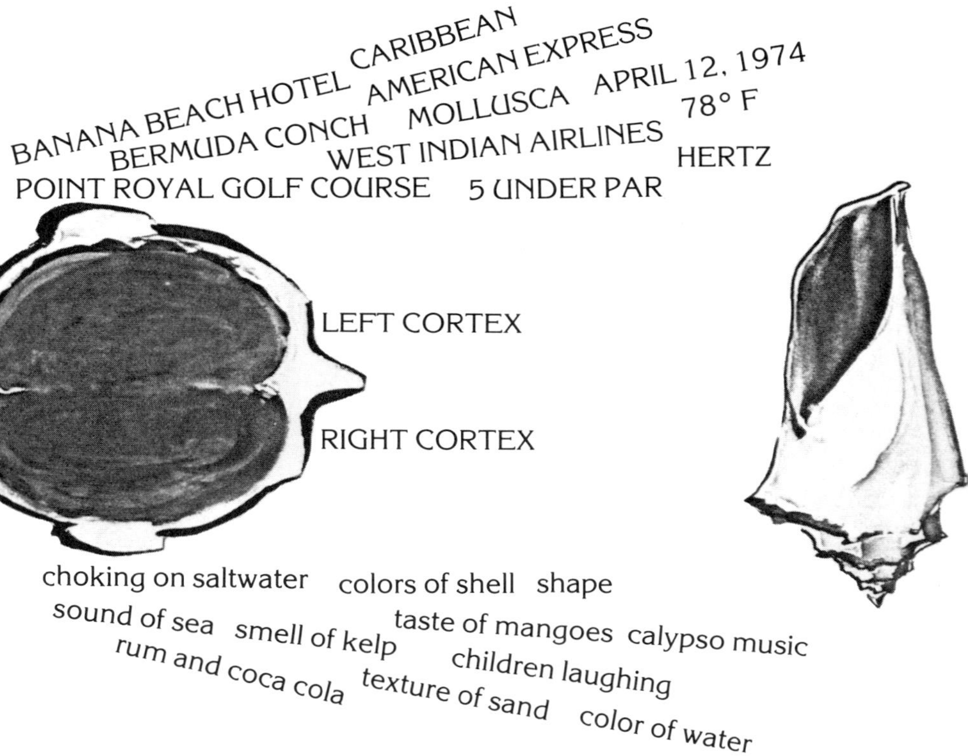

Figure III

This is a portrayal of the way the brain/mind experiences the more holistic context of "shellness." The brain/mind cannot experience a single thing; instead, an ecology of experience accompanies each event.

Be assured that Figure III is diagrammatic; the location of the variety of reductive and metaphoric experiences is strictly graphic. Even so, the primary message is linked to the network of interrelatedness that attends any experience. *One cannot experience one thing*. Instead, each life experience encodes a whole ecology of perceptions, conceptions, and metaphoric possibilities inside the configurations of memory.

Sir Charles Sherrington's poignant metaphor of the "enchanted loom" serves us well here. What we experience and learn is not a single point in our consciousness. It is, instead, rather like a fabric, woven of many threads, of many colors, and of many textures. The "enchanted loom", though richer than most, is still a closed-system metaphor. It, like the hologram, is a physical expression of processes that are open. In an earlier work, *The Metaphoric Mind*, I warned of the use of closed-system, physical metaphors in the explanation of open systems and open-system processes. The brain-mind system is a natural, open system; thus metaphors like looms and holograms are tacitly limiting. This is also true of computers, gears, switches, cylinders and clockworks, all of which I have heard used to explain mind function. The mind is holonomic; such metaphors are not.

In the holonomic mind, the emphasis is on the whole . . . the entire context of a thought. Central to the basis of this model is the assumption that the whole brain (not just the cortex) is excited during each experience. Further, it assumes that the central nervous system and the entire body are also party to the experience. In a sense, each experience creates a particular and unique neuroelectrical field. There is a response by those parts of us that operate under the domain of virtual consciousness . . . our body chemicals.

The primary body chemicals of concern here are our neurochemicals. Responding in the domain of their own virtual consciousness, billions of atoms, ions, and molecules shift their orientation to achieve a higher kind of harmony with the neural energy field. This happens not in one place, but across the entire neural energy body. As surely as the substance of the planet once wove and moved through space to enjoy stability in the cosmos, the effect of thinking a thought reinvents the mechanisms of all our beginnings.

A series of experiments in the early 1800s by a European physicist named Ernst Chladni sheds some light on this issue. Chladni fastened large metal plates onto sturdy posts driven into the earth. The plates were strong but flexible. The supporting posts attached to the center of each plate's underside made the system quite firm. On the surface of the metal plate, Chladni sprinkled finely divided materials like cork dust, sand, metal flakes, and poppy seeds. He then drew a violin bow down across the edge of the plate. Naturally, the whole system began to vibrate.

The vibrations caused a great turmoil in whichever substance he had sprinkled on the plate. After he stopped "playing" the plate with the bow, there was always a distinct and symmetrical pattern made by the bits of material on the top of the plate. What happened was clear. The stroking of the plate with the bow set up vibrations in the plate that caused the particles to respond. In accordance with their mass and the pattern of vibrations, they would migrate from region to region on the plate. Once they found a place in which they experienced resonant harmony, they stopped. Since some of the portions of the plate were too "energetic," the particles couldn't rest there. Instead they moved on. Eventually, they synchronized with the plate.

This is analogous to neurochemicals migrating about the brain to achieve an equilibrium of mass and charge in response to neuroelectrical fields generated in thought. Ralph Abraham terms the configuration of such processes as macrons.[4] A macron is the configuration that emerges when any material substance is in resonant harmony with the energy field it is experiencing.

Abraham identifies four main classes of macrons. These are physical solids (similar to the Chladni figures), physical isotrope liquid, physical liquid crystal, and physical gaseous. All of these are reasonably well explored and reasonably understood. Two more kinds of macrons are as yet quite embryonic in terms of understanding. These are chemical and electrical macrons.

Abraham is highly cautious about electrical macrons in particular. Yet his speculations blend beautifully with the holonomic model. He writes in *The Evolution of Consciousness*:

> The description of basic electrical macrons is included here for the sake of completeness—in spite of being based almost entirely on speculation—and because of my belief that it will figure vitally in the understanding of the brain.[5]

Abraham continues and postulates a concept of mind that is wholly consistent with the holonomic model developed here. He suggests that a thought is a macron of the bioplasm of the brain. In other words, a thought is a three-dimensional configuration in which neurochemicals achieve a resonant harmony with energy fields created by the brain. He further suggests that temporary macron formation is the common way of thought. The temporariness results in short-term memory. If a particular macron, however, is repeated over and over again, then some of the affected neurochemicals fix on brain tissue and provide a kind of permanent neurochemical halo. The result of this would be easily regenerated macrons or . . . long-term memory.

The primary relationship between the holographic ideas of Pribram and Bohm and the macronic options of Abraham is that the union of both provides us with a workable physical metaphor for both structure and process in mind function. If the brain is structurally a holographic entity with all facets of experience encoded throughout itself, then the process of generating a macron excites the brain in intertwining networks of stimuli.

Such a model offers much in terms of learning and recall. Brain-damaged patients often show highly differential recovery patterns. Neurosurgeons have long noted that the recovery of many patients with similar surgery was remark-

ably different. For example, patients with left cortical surgery or left hemisphere strokes often suffer an impairment in their logical, linear, sequential modes of knowing. Most often this is manifested in a loss in the ability to speak or read. To an adult, this is fiendishly frustrating and discomforting. Yet recovery is miraculously possible. The primary difference in the rate and completeness of the recovery has seemed to depend on the approach of the recovery therapist. Those who focused upon the ways of knowing that were *lost* had tedious and tortuous paths to follow. The therapists who celebrated the modes of knowing that were not impaired saw remarkably rapid rates of recovery in their clients. Figure IV illustrates this:

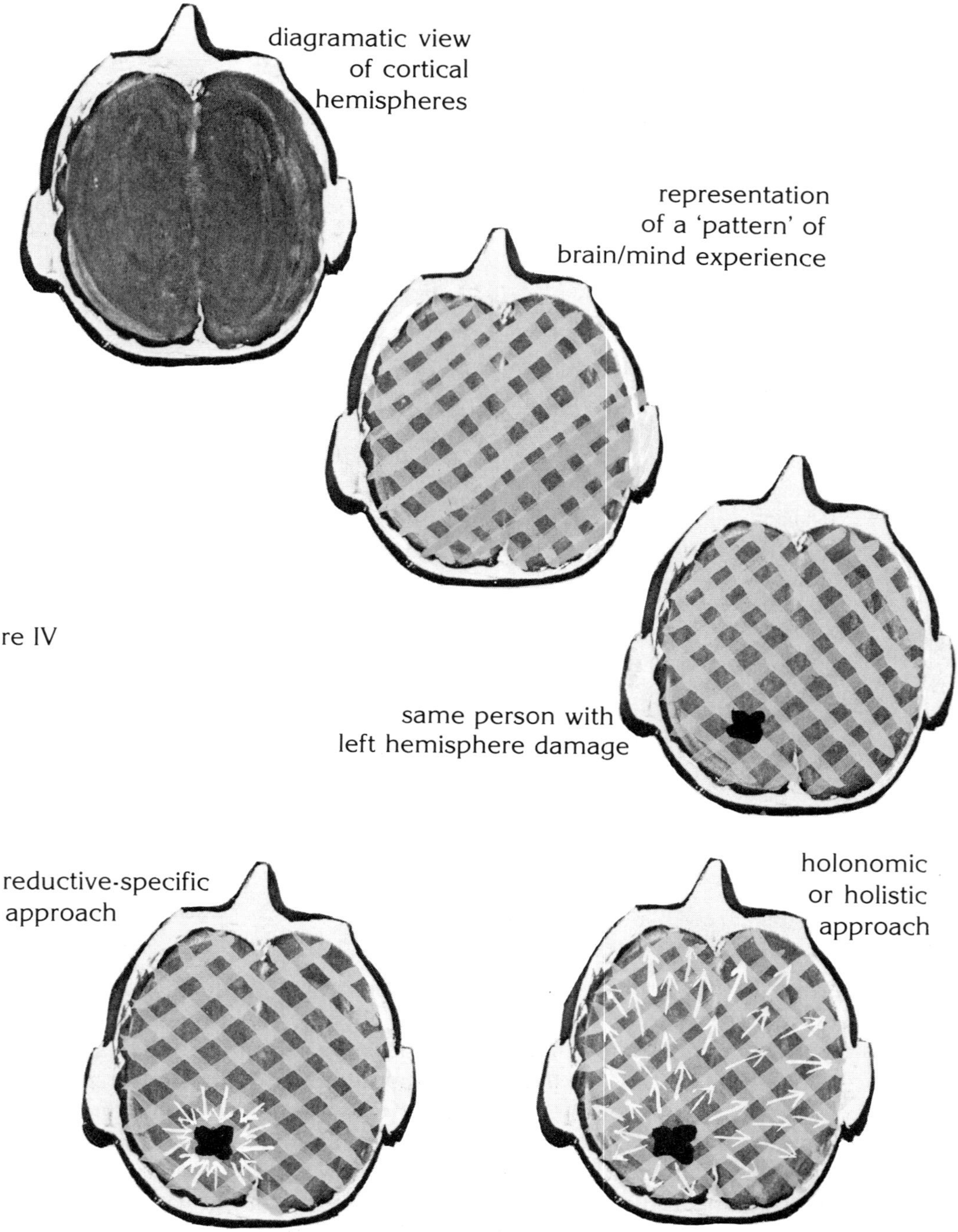
diagramatic view
of cortical
hemispheres
representation
of a 'pattern' of
brain/mind experience
same person with
left hemisphere damage
reductive-specific
approach
holonomic
or holistic
approach
focus on the functions that were lost
focus is on the whole pattern

Figure IV

In education the same results have been achieved without the trauma of surgery or stroke being manifested. Instead, those malevolent neural gremlins called "learning disorders" were at fault. Children who have difficulties in reading and writing are subject to the same two approaches as those people depicted in Figure IV.

If the disability is focused upon, improvement is slow and often accompanied by a negative attitude toward the skill being developed. Conversely, if the child's abilities are celebrated, recovery is approached from a base of strength rather than weakness.

The holonomic model, then, offers us a much more holistic vision than we have been previously afforded. It allows us as teachers, therapists, and philosophers far more freedom than we have had before. Strangely, it returns us to an open-system view of mind. It affords us the realization that there is no parapsychology, no extrasensory perception. These functions exist, but are not separate. We are provided with the realization that anything the mind can do is normal.

To return again to Bohm's words:

> The true state of affairs of the material world is wholeness. If we are fragmented, we must blame it on ourselves.[6]

With an open system mind, a holonomic mind, becoming our basic model, then it is clear that the mind and universe are one. Perhaps when our mind performs in ways that transcend space and time we are simply wedding our consciousness with larger, more pervasive systems of knowing. Chapter 4 explores some of the possible mechanisms of such linkages and some of the effects.

Minds are limited because our concept of mind is limited. Consciousness is narrow because in listening to the closed-system thought of our technological cultures, we narrow our attention. Again, consciousness is what we pay attention to. In the past, our view of brain, mind, and consciousness was defined by a narrow, reductive mode of knowing. As functional as those modes have been in the past, we are witnessing trends toward a larger vision. As surely as the paradigm of reductivism is shifting toward holism, so too are we honoring our origins. We will not abandon the specific wisdom of closed systems, but . . . we will no longer be imprisoned by them.

4

OPEN MIND

The holonomic model of mind is a model of an open system. It is open in the sense that the brain appears to function as an open system. Historically the brain, like other body organs, has been considered closed and "internal." Yet when the brain performs its orchestration of the senses, of perception, and of thought, it weds itself to the cosmos. It creates thoughts in ways that excite its holonomic wholeness. The boundaries of mind extend far beyond the organ of the brain, far beyond the entire body.

It is *consciousness* that limits the range of attention we honor in scanning the millions of contexts created by the brain-mind system. It is as though thoughts created by the brain-mind originate from endless cerebral meadows rich with flowers, butterflies, and a myriad of other life forms. As we walk such meadows, our path is always narrow. What we become aware of is always limited and highly selective. It is consciousness that makes that selection.

Consciousness is what we pay attention to. In the meadows of mind we may one day gather a bouquet . . . another day observe insects . . . and yet another time map the paths we have taken. But only rarely do we experience the meadow in its entirety. For most of us, "paying attention" in consciousness is a process of reducing our vision. It is an act that involves a narrowing of perception.

I once asked a group of junior high school students to map the school campus. Immediately the concept of "mapping" began to limit their consciousness. They set about gathering rulers, tape measures, compasses, pencils, paper, and other paraphernalia classically associated with mapping. I watched them for a while and saw the mountain of mapping gear pile up. Then I stopped them. "You don't need all that," I said. "Today you are going to map how the campus feels to the touch."

They stopped. After a few remarks of incredulity, they became quietly attentive. "Come with me," I said, picking up a paper sack I had prepared. I led them outside. It was late spring, and the Colorado earth was soft and warm. I asked the students to divide up into pairs. At that point, I disclosed the contents of the paper sack. Inside were blindfolds. One partner from each pair of students then took a blindfold from the sack and prepared to blindfold his or her companion. I

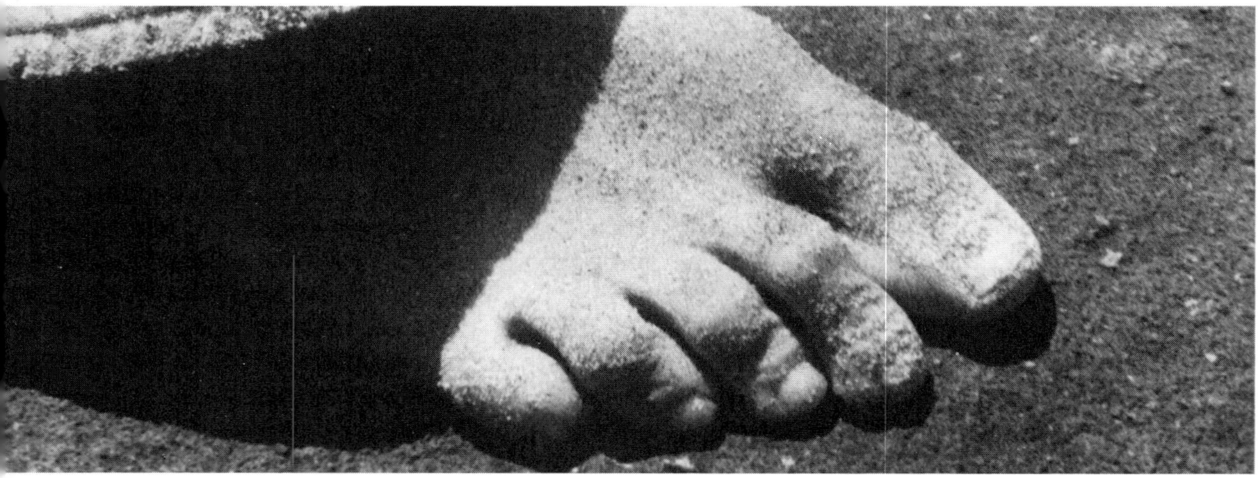

stopped the students at this point, and asked that those students who were to be blindfolded take off their shoes and socks. Enough of the early childhood glee still lived in these boisterous twelve-year olds, and they took off their shoes and socks with great and noisy dispatch.

Then the blindfolds were fixed. The sighted "buddy" was responsible for recording what the "tactile" student experienced. Also, the sighted student had to be sure to help avoid injury on the part of the blindfolded student. At first, this activity was noisy and filled with trepidation. Gradually, it got noticeably quiet. The tactile students began probing the earth with their toes and feet in ways that gathered in the maximum information. In later discussions, nearly all confessed that it would never have occurred to them to map that way—nor would they have been able to predict how much they were able to sense in this way.

These students simply experienced what some might call an altered state of consciousness. Charles Tart, the noted researcher and author in the field of consciousness, says that normal consciousness is the one in which people spend the major part of their waking hours. Tart defines an altered state of consciousness as "one in which (an individual) clearly feels a *qualitative* shift in his pattern of mental functioning."[1] He suggests this is not more of the same kind of consciousness that one ordinarily experiences, but is rather a true shift in the way of experiencing "reality."

Tart clearly is uneasy about the word "normal." He is quite clear about how hazardous it is to pretend to know what is normal. It is obvious that the San Francisco stockbroker and the Kalahari hunter mentioned before have very different realms of "normal" consciousness. If one were suddenly to experience the consciousness of the other, this would clearly be considered an *altered* state in terms of the prior definition.

Yet as I have explored the models of mind, I have become more convinced that anything the mind can do is normal. If the mind is an open system, as the holonomic model suggests, then its function is affected by internal and external systems of energy. The macrons we discussed in Chapter Three assist us here.

Assume first that when the brain creates a thought, it also creates a macron. Further recall that a macron is a patterned configuration that results when material substance achieves resonant harmony with the energy it experiences. It is also necessary to recall that this energy/material configuration encompasses the whole brain and perhaps the entire body. If all these conditions are assumed to be valid, then normal consciousness is made up of those portions of the macron/thought that we reflexively pay attention to. Thus "normal" consciousness is an expression of that which we *reflexively* pay attention to.

Think of consciousness as a ray of sunlight lancing through an opening in a cloudy sky and pouring out onto the earth. This ray of finite width scans only a small portion of the landscape. So, too, consciousness scans only a small portion of our mindscape. Moreover, it is likely that self-imposed "habits" in consciousness create a narrow range of options in terms of "normal" thought. Continuing with this metaphor, an "altered state of consciousness" results when another opening appears through the clouds and the sunlight strikes a new domain within our mindscape. The second opening is distant from our "normal" opening and thus illuminates terrain never before disclosed.

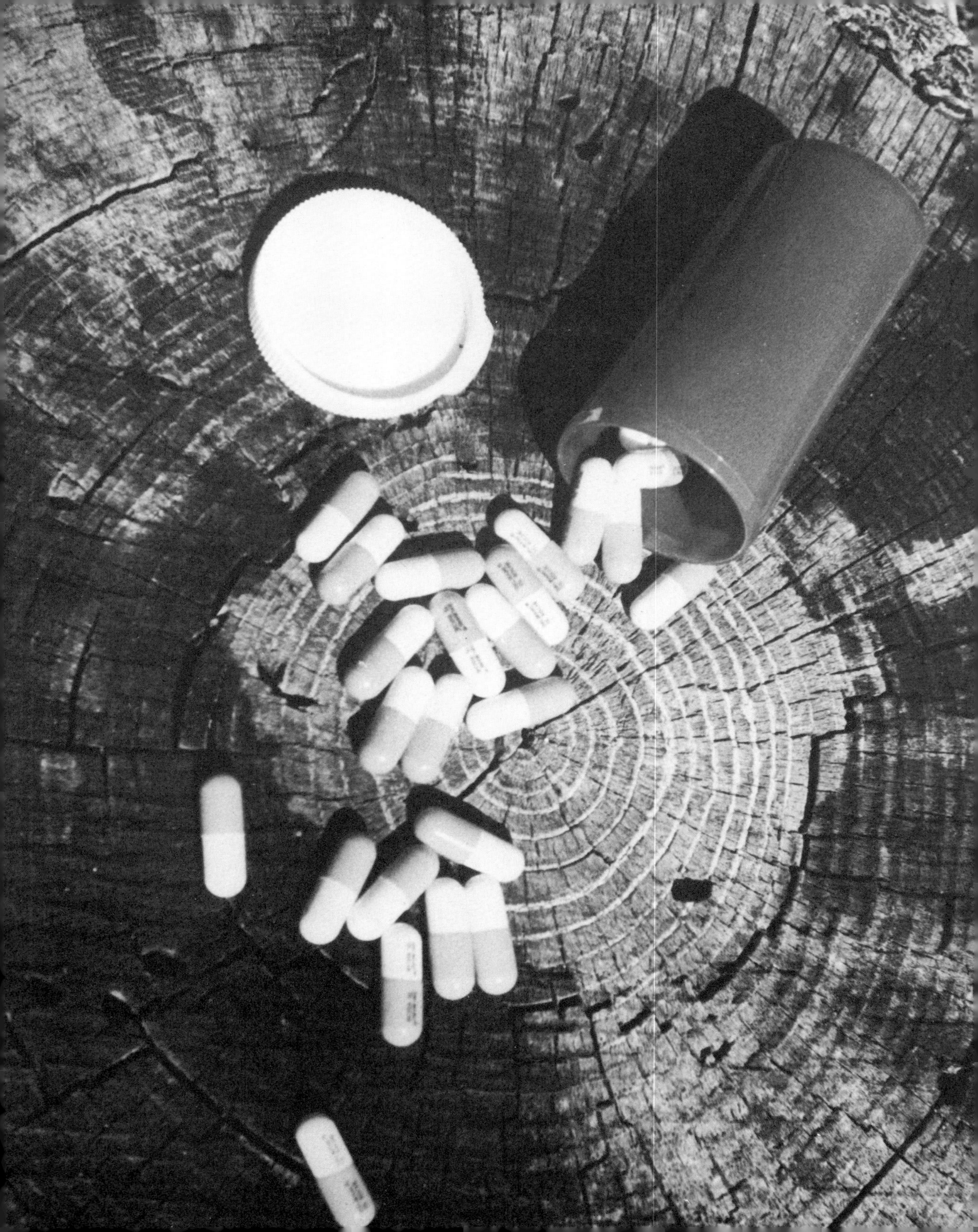

What bothers me about this metaphor is the notion that exceptional consciousness is accidental (subject to the caprice of holes in the clouds). It suggests that we might have to *do* something to our mind system to experience new consciousness. For the last several decades, numbers of people seem to have thought that an appropriate thing to do to the mind system was to introduce a large array of alien molecules in the form of drugs.

Hallucinogens, alcohol, narcotics, barbituates, etc., have become the ways to seed the clouds of mind and to allow more "light" to come in. In a sense, these substances justified the use of the word "altered" state of consciousness. In our chemically-addicted society, the introduction of long lazy molecules, or short aggressive ones, to the macrons of mind could perform acts of alteration requiring little intrinsic mental activity.

Consider this: If internal energy/material configurations—macrons—set the "normal" mindscapes upon which consciousness can focus, then the introduction of new material (new molecules, atoms, and ions) will change the shape and pattern of the macron. After all, the shape and form of the macron results from a resonant harmony between energy and material. Add new material and keep the energy the same and a new pattern will emerge.

In response to all of this, I prefer the use of the phrase *extended state of consciousness*. Extended as opposed to altered. I have no quarrel with the notion that external substances do alter consciousness. But I am troubled by the implication that such acts of ingestion are necessary in order to widen and extend the realm of what a person pays attention to. All this is summed up by the belief that if consciousness is what we pay attention to, then *we can extend consciousness by paying attention to different things*. It probably means that I am more interested in the energy part of a macron than I am in the material substance with which it establishes harmony. I am inclined to retain the integrity of the natural brain, free of external drugs, as I explore the ways of extending consciousness.

It is clear that many neuroscientists are now considering the brain as the largest gland in the body. They cite the vast array of self-generated substances the brain manufactures. It is with these internally-derived substances—the substances of evolution which have been provided by the planet and manifest in the brain—that I prefer to seek new contexts of neuroenergy generation. For me, the most appropriate way to serve the substantive side of planetary consciousness is through nutrition and self-generated flexibility in consciousness, rather than through the use of artificial additives to the mind.

As I use the term, "altered consciousness" requires the input of external *material* into the brain/body context. So far, I have said little about external *energy* input, and particularly such energy input into the open-system model of the holonomic mind. Much of what I have said in previous chapters has had to do with *internally*-generated energy related to thought. We have focused on how one's brain, in a self-initiated aggressive mode, creates thought/macrons from within. A powerful question arises: Is a brain susceptible to external inputs of energy when in receptive, yielding, and passive states? Can external energy create thought/macrons that extend consciousness and enrich normal consciousness? Yes.

Anyone who has paid attention to night dreaming is generally convinced of our susceptibility to external energy configurations. In night dreams, nearly everyone experiences *deja vu*, *prescience*, and *precognition*.

Deja vu may be understood as a circumstance in which a presumably new experience is greeted with an instantaneous rush of familiarity; it seems to have happened before, with no basis in known experience. *Prescience* is foreknowledge; knowing that something is going to happen, before it does, and often in some detail. *Precognition* may be understood as understanding or knowlege that emerges without basis in prior known experience and information.

If one pays attention to dreams, it is clear that the precursor to a later, waking experience is often expressed in a dream. For example, I remember the account of a waking, *deja vu* experience centering around a staff meeting, shared by a person in a dream seminar that my colleagues and I were conducting, at our annual Solstice Seminars.

The woman who had this experience had been in the midst of a typically busy day. As part of her professional responsibilities, she began, somewhat routinely, preparing to leave her office to attend what she knew would be an important staff meeting. For a moment, she felt drenched in a sense of bewilderment. Stopping at the door of her office, she felt a chill of fear. *She knew that she had experienced this moment before*. She closed her eyes, and consciously tried to gather herself together. Her mind filled with images of everything and everyone who would be waiting at the meeting. She could see who was there, where they were positioned in the room, what they were wearing, and their attitudes in conversation. She could even see the agenda topics, despite the fact she had not seen or been told them in advance. She opened her eyes, shook the images and the feelings, and walked briskly on to the meeting. There, she opened the door . . . and it was all just as she had seen it, complete with the mid-meeting flood of coffee dumped by the associate director of the staff onto a stack of critically-important papers.

The dream seminar legitimized this previously inexplicable experience for the woman. In fact, she began paying attention to her night dreams in some detail. Within a matter of days, she was able to recall her dreams in great detail, including dreams she had had in the past. In the course of this process, she came to the realization that she had dreamed about the staff meeting some days in advance of its occurrence. The exactness of her dream images in contrast with the actual meeting astounded her. Once again, as she reminisced and re-constructed the dream and the meeting, she became uneasy.

We assured her that her uneasiness was a result of her own biases about what was "normal." As the seminar continued, she was able to link many of her *deja vu* experiences with prescient dreams. Prescience, again, is having knowledge of events before they happen. Literally hundreds, if not thousands, of incidents such as this one appear in the literature of dream research and the study of so-called paranormal phenomena.

The research also indicates that dream *transmission* is possible. Stanley Krippner has been involved in a series of studies that indicate the ability of the mind to transmit and receive images in dreams as a demonstrable possibility.[2] My own work has indicated sharing of dreams among people with a strong sense of bonding as a common phenomenon.

How does this happen? Do the people involved have tendencies toward extended consciousness? My response: the experiences are normal. Actively *or* passively configured thought/macrons are as active in sleep as they are in waking life. I believe the mind is completely capable of accepting incoming energy, and thus creating contexts of thought that can be scanned by intrinsic consciousness.

In the spectacular, such experiences have been called "revelations." In Western and Eastern cultures, they are reserved for special use by spiritual leaders. Native American Indian people and Australian aborigines, as well as most people of native cultures of the southern hemisphere , consider such experiences absolutely normal to everyone. Some of the tribal people develop more grace with vision and dream and thus achieve special status, but the primary form of consciousness is afforded to all.

The holonomic model of mind includes the potential for external energy configurations, as well as internally-originated configurations. Thus, it is normal for both intrinsic and extrinsic energy to be appropriate in the creation of thought. People in our particular dominant culture are comfortable with external influences on the sculpturing of thought/macrons as long as they are delivered during wakefulness and through appropriate channels. In fact, those who readily submit to external structuring of thought are called students. Paul Brandwein calls the process "schooling."[3]

Yet if someone admits to having "learned" or been "informed" from some nonconventional source, the time for persecution begins. In our culture, it is safer to narrow consciousness than to extend it. It is more appropriate to stay with the modes of knowing that have been institutionalized than to create or accept new ones from without. A person who survives persecution for holonomic knowing and persists in acknowledging this wisdom is called a psychic, clairvoyant, mystic, medium, visionary, or perhaps hypnotist. To most who conform to extrinsic definitions of normalcy and standardized versions of consciousness, these "strange thinkers" are merely the insane who are not dangerous enough to be locked up.

And yet it is accurate to say that many such open-system thinkers have had to resort to a charade-like life. Bizarre and flamboyant behaviors have been affected by many who find such actions the only ways in which to cope. They begin to act in ways that will excuse their thought patterns. To many others, the vast majority of experiences in extended consciousness are simply denied. In the face of the oppressive presence of the myopic and narrow definition of normalcy honored in our culture, any wider view of normalcy is denied.

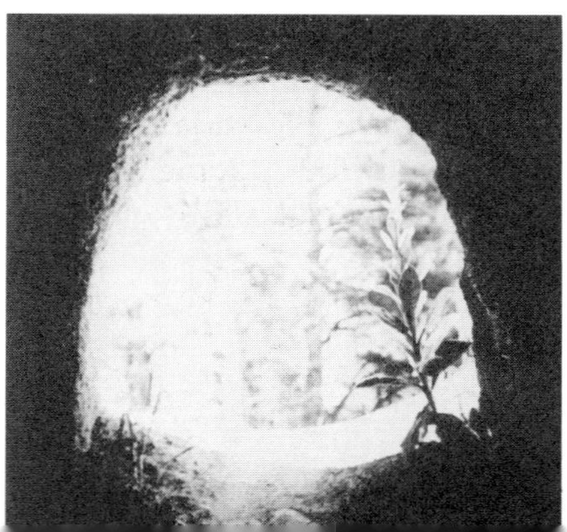

To parody Plato's cave . . . it's as though we stand inside the threshold of a technologically-closed dwelling. Adolescent in our conception of mind, we stand trying to decide whether to go back into the safety of the box, created and compartmentalized by the culturally-conforming mind—or to venture out into the open system of the planetary mind. We are struggling with the comfort of a closed-system image of the mind, while we sense the excitement of the possibilities of an open mind. Our self-made images of mind have, in the past, entertained metaphors of boxes, circuit boards and computers . . . rather than feeling the open-system possibilities of the mind as music, clouds or wind.

Is the promise of the holonomic vision too ambiguous to entertain? Is it so liberating that we are unable to resist? Are we ready for the paranormal to be normal? Are we ready to admit that visions and vibes, telekinesia and telepathy are all real and normal expressions of the holonomic mind?

The notion that thought is energy in harmony with matter lets us look anew on the idea of synchronicity. To become synchronous, one directs oneself to resonate with each experience. Harmony—resonant harmony—is synchronicity. If thought is in fact an expression of resonant harmony, then it can be a true model of the coming demands upon human culture and upon natural systems. To think a thought can be an invitation to our consciousness to wed the brain, mind, and body into harmony . . . into entrainment. Entrainment is a kind of harmony that develops between two or more systems that are in each other's presence. Two or more pendulums swinging in a non-harmonic pattern gradually and mysteriously adjust so they become synchronized and begin to swing harmoniously. Once we extend our consciousness toward this end, we find ourselves again in harmony with the planet.

In his book, *The Silent Pulse*, George Leonard summarizes this concept when he claims, "Our ability to *have* a world depends upon our ability to entrain with it."[4] Leonard reports the expression of a young man who experienced this kind of unity. The seventeen-year old writes:

> Even though physically separate, I knew a tree, grains of sand, sea, flying birds. Everything was God, holy; as God is total, so the driftwood branch was holy. This must be the stuff religion is made of. Never before or after have I felt so alive.[5]

Perhaps the human has the capacity to be god-like. But if all the images that all the cultures have of God are images of oneness and of unity of omniform presence, then we cannot seek godlikeness in closed systems. We are children of culture as surely as we are children of nature. Culture is the most recent in our experience. But the Mother, in silence, has given of her body to create our most recent commitments to culture. She has survived our compulsion toward closure . . . but her wounds are many. Perhaps we have found ourselves ungodly long enough. Perhaps we are inviting our consciousness out of confinement and back into a system as open as the cosmos.

5

HOLONOMY

Perhaps the most intriguing aspect of the holonomic model is that it is not limited solely to explaining brain function. We seem to have a model that incorporates, without detailing, all of the concerns voiced in modern physics about matter and energy. Questions about the laws of thermodynamics and entropy are afforded within the holonomic model. The wisdom of the West, including Newtonian physics and Newtonian thought patterns, is persistently being transformed.

It is clear that physicist David Bohm has extended the holonomic model to include the entire universe. Bohm feels that the brain, the mind, and in fact all material in the universe are part of a single, larger scheme of interference patterns in the frequency domain.

Recall from Chapter 3 that the hologram is not a picture as such. It is an image that results from the interference of wave energy of different frequencies. Bohm, as well as a number of other physicists, believes matter—the substance of the universe—is also an expression of different frequencies. Matter then is no longer the solid, predictable stuff we learned about in school. It is, rather, an aggregate of frequencies. It is the ghost music of the domain of frequency.

It is an assault upon traditional reason for most of us to try to create metaphors for the frequency domain. First of all what does it mean? Much of our training in science, both technically and in a popular sense, includes choices of metaphors of *matter*, of solid substance, to express the ways the world is put together. Most of us found comfort in the solid-liquid-gas triad. Each of these were manifestations of the other. It was just that the linkage between the material differed.

A solid existed when atoms or molecules clung tenaciously to each other and formed a reasonably non-yielding substance. It could be bent and broken, but it possessed a physical integrity. In a liquid, the matter had a casual relationship with its own material structure and the substance would slip and slide in a manner showing less integrity. And a gas? . . . a gas is so detached that it possesses almost no integrity. The atoms and molecules act almost independently of structure. A gas is like a solid postponed.

Thus, stability became the most pervasive expression of matter. Its "solid" and "fixed" qualities crept as innuendo into the metaphors of human existence. Solid as a stone, sturdy as an oak . . . all expressed more integrity than liquids, which were slippery, fluid, wishy-washy. Gases were so ephemeral that they were represented by the most evasive metaphors.

But in regard to holonomy, these metaphors of solidness, when ascribed to matter, lead us to a kind of closed-system thought. Atoms are *not* solid. Neither are any of their parts. They are in reality expressions of vibratory energy. The particular patterning of each of these vibrations possesses a particular frequency. The frequency is the expression of how often the vibrations occur in a specific time. So all of the different particles that make up atoms turn out to be specific expressions of patterned energy, vibrating in specific frequencies in this domain of constant change.

Stability in this world is more of the nature of subatomic music than of structure. It is a world of balanced harmony that changes, resonates, and configures the patterns of basic stuff of the universe . . . the patterns we call matter.

Bohm suggests that matter is in fact a holographic expression of these forms of wave energies, and it appears throughout the universe where the conditions are right. Bohm likens the process to the way clouds appear and disappear in the atmosphere. When the energies expressed as temperature, pressure, and moisture are just right, clouds appear. If those conditions change, the clouds slip back into the atmosphere as dispersed water droplets. In Bohm's words, they "fold" back in.

What this means is that regardless of how "closed" and substantive a system really appears to be, it is truly an element of a larger, "open" system. The physicists who once thought of matter as being little balls of positive and negative charges are now growingly convinced that matter is patterned energy. Patterned energy is a far more open model than solid spheres.

Everywhere that one seeks closedness, the prevailing benevolence of openness persists. And so too it could be with people. Yet with people it is clear that although the possibility for openness exists in the natural, physical construction of humans, it seldom prevails.

Let us pursue, for a moment, the natural construction of humans. In holonomic perspective, humans—like all other objects—are specifically configured patterns of energy/matter temporarily stable in the universe. The period of stability is called life. And the point at which human energy/matter begins to change its configuration and begins the process of reuniting with the cosmos we call death.

For many, the closed-system comfort of matter being solid and material substances being permanent is nearly impossible to give up in our considerations of life and death. For centuries, life has been treated in the linear way; it assumes a substantive beginning at birth, a substantive maturation and growth throughout life, and a dissolution of that substance with death. Much of the history of humankind has been centered around ascribing material significance and substantive meaning to the period between birth and death. All cultures on the planet have built rituals and rites to ascend and descend through the "material" plane. The formalization of these rituals and rites is called religion.

From a holonomic perspective, the entire concept of birth, life, and death, is not a symphony unto itself, but rather it is a single movement in an infinite pattern of cosmic music extending from and toward infinity. Such a perspective must unsettle the closed-system thinkers. It denies the separateness that so nurtures the closed mind. If atoms cannot be solid and of substance, then does this mean *nothing* has meaning? Not at all. It means that *everything* has meaning. The human as an expression of an open-system, holonomic being is a much larger and more integrated expression of the universe than the finite, closed-system born-live-die human.

Even in the context of the spirit and spirituality, holonomy testifies to a large unity. In most cultural religions, the body—the material being—is lost, but the spirit prevails. In holonomy, *the concept of loss does not exist. All* of humans, *all of all,* is bound toward continuity. New patterns emerge from the old . . . nothing is interrupted.

Holonomy and spirituality are probably intimately related. Perhaps even identical. Consider this definition of spirituality. *Spirituality is that quality of being that expresses the bonding of all living and non-living things into an evolutionary unity.* This differs from conventional western religious doctrine in that for most spirituality and religiosity have been considered identical. In such a religious view, that which agreed with and conformed to doctrine had evolutionary continuity and that which didn't conform was destined to decay . . . dust unto dust. Matter . . . material substance, in this view, is not part of unity. In the view of spirituality held in this book there is no such separation.

By whatever means each of us describes the assembling of non-living energy pattern into living energy pattern—whether thermodynamics, genetics, and/or God—the stuff of which we are made is still in union with all the substance of the universe. In the perspective of holonomy, this unity is basic. As each bit of matter/energy begins, evolves, and is transformed, it still retains its unity with the cosmos. In this sense, the way we stand—and have stood historically—in awe at the miracle of birth and life, we may now extend to the rhythm and flow of life back into the more metaphorical mother, an evolution into what is popularly called death.

Like Bohm's metaphor of clouds, the matter/energy configuration of humans is also part of larger cycles of rhythmic being. It is clear there is enough stability to these patterns—whether clouds, lilacs, or humans—for some continuity to persist. Though the detailing is not clear, the whole concept of holonomics is neither a voice for science nor for religion. Nor does it negate either. For both are simply politicized versions of kinds of spirituality. Science has politicized the spirituality of logic . . . Religion has politicized the spirituality of mysticism.

Within the workings of mind, the perceptions that we formalize into fact and truth are in reality only possibilities. Each single thing that each human "knows" is only a single possibility in a vast array of possibilities. The physicists

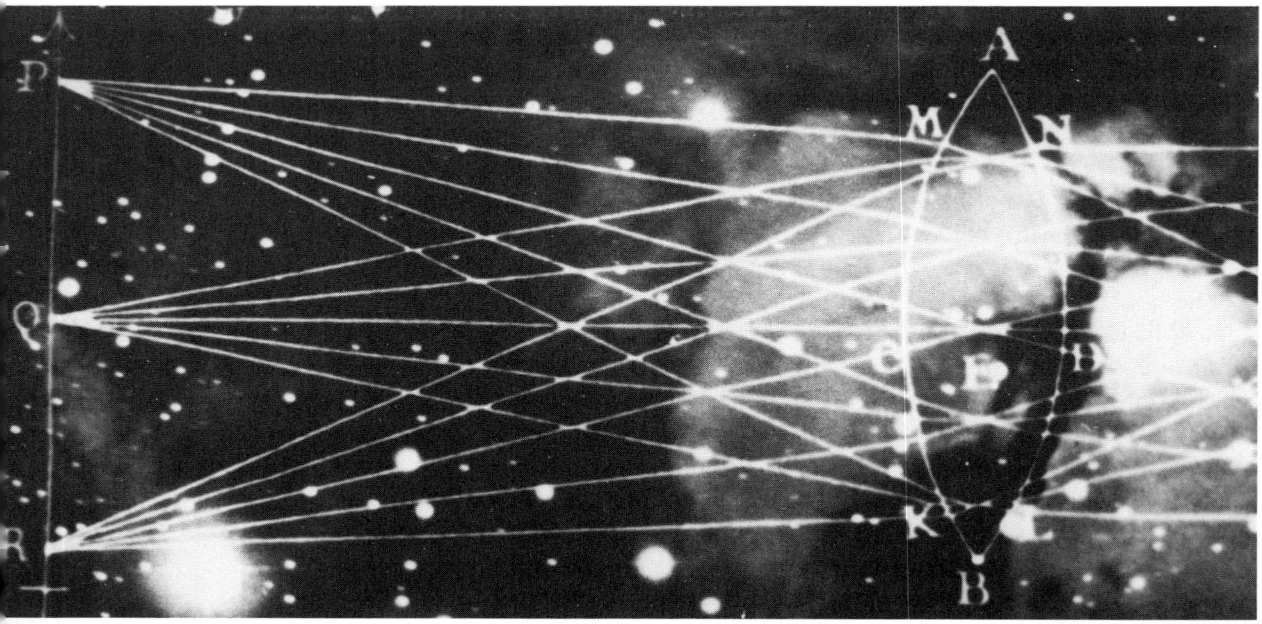

are clear on this point as they pursue "objective reality" with all the ferocity that our grand, technological age can arm them with. But as this pursuit goes on, it becomes more and more clear that what used to be true and fixed is now only probable and transitory.

Once beyond the safe world of pulleys, levers, and apples falling from trees, the certainty of physics transforms to more mystical qualities. Unfortunately for the technological, Western-mind view, reality was popularized as being linked to the world of certainty in the physics of Isaac Newton. Fact and truth were described and defined against the tapestry of mechanical certainty born of these same pulleys, levers, and falling objects mentioned above.

Albert Einstein's immense contributions in Western physics crushed the safe "realities" of Newton into a narrow range of cases when he introduced into the arenas of reason the unity of the infinitely small world of the atom and the infinitely large world of the universe. It was clear that the Newtonian world of "our size" was filled with biases about certainty that led most humans into a false vision of reality. The Western version of the human mind began to favor forms of consciousness that were filled with fixity and closedness. It became increasingly difficult for human consciousness to entertain the following statement by Fritjof Capra from his book *The Tao of Physics*:

> In the Eastern view, the reality underlying all phenomena is beyond all forms and defies all description and specification. It is therefore often said to be formless, empty, or void. But this emptiness is not to be taken for mere nothingness. It is on the contrary the essence of all forms and the source of all life. [1]

To humans who perceive themselves as closed-system beings, separate from the planet and its other inhabitants, the quote above is filled with dreadful ambiguities and nonsense. To summarize, if I perceive myself as a reality bound by birth and death, and as a closed-system element in a closed-system world, then the whole idea of holonomy might appear a rather repugnant fantasy. So too would the many possibilities of a holonomic vision remain invisible to one blinded by a persistent search for closed-system realities.

All the ideas cited above represent a blending of those who are clearly uniting in the cosmos. The emerging vision of unity is perhaps best expressed by British scientist Rupert Sheldrake[2]. Sheldrake has offered a remarkably open and all-inclusive version of unity in the universe. He believes that the universe is governed not only by the forces discussed in Chapter 1, but also by organizing fields generated by the interaction of these forces. He calls this invisible organizing matrix the *morphogenetic field*. *Morpho* means form and *genetic* expresses coming into being.

What this means is that whenever energy is patterned into matter or vice versa a morphogenetic field is created that resonates throughout the cosmos. This resonating field then guides all future patterns of organization. Sheldrake may well be postulating the scientific expression of the "consciousness" of the cosmos which is central to this book.

Where then has this excursion into holonomy taken us if we are to return to the concept of the person? How does the posing of the open-system human relate to the context of human personality and the human role as both individual and member of a culture of humans?

One of the possibilities most overlooked by closed-system thinkers is that open systems exist. In our culture—our Western, technocratic culture—each human is subjected to a non-ending array of invitations toward fixedness, certainty, and closedness. Nearly all the effort of the intellect in both tacit and formal training is aimed toward becoming more elegantly closed. In honesty, the more highly educated were at one time those who were more sophisticated in their closedness. In a sense, the highly educated person was an authority whose closedness was based on years of gathering reasons for closedness. The ignorant were simply closed without reason.

At first, it may seem I am arguing that only Western technological cultures nurture such closed-system thought. Not so. This is a cultural phenomenon. All cultures on the planet establish contexts of closedness, and these contexts are the basic elements that provide identity and distinctiveness to that culture.

Eastern philosophies, for example, stressed the interior of the person and nurtured a search for meaning in the center of existence. This contrasts sharply with the Western view which stressed a search for meaning in the external. Rene' Dubos in his book, *Wooing the Earth*, points out that Taoists emphasize the unity and rhythm that pervades the universe while Shintoists deify nature.[3] Yet both tend to translate these into more precise cultural lifestyles as opposed to the establishment of harmony with nature. Nature is central to them only as intellectual exercise. The history of both East and West is clear in testifying to the exploitation of nature, justified by world view. Neither view is open system. Neither view is truly holonomic.

But then, can any culture be open-system? . . . Can any culture achieve visions that extend clearly beyond the "pattern" used to describe the belief systems of the culture?

It is clear that when one does make cultural comparisons, there is evidence that although closed-system thought is necessary to establish the essence of a culture, many cultures also engage in varying degrees of open-system allegiance. Navajos and other Native American Indian tribes have a very large proportion of their thought devoted to open-system approaches. Nonetheless, the Navajo and their neighboring Hopi are sufficiently closed in their thinking to easily differentiate one from the other.

Several times in the previous pages I have mentioned distinctions between Eastern and Western world views. A distinction between Eastern and Western world views is grist for popularization in contemporary consciousness. While I believe the "revolution" of popularized enlightenment expressed by the

Western discovery of Zen Buddhism, Hindu literature, and Sufism is worthwhile, I still view it as incomplete.

Although Eastern and Western world views are complementary, I suspect that their synthesis will still result in a partial vacuum of consciousness. Comparisons of East and West represent comparisons of *cultural* perspectives. They represent the complementarity of one-half of a perspective. This means another source is needed . . . and, for me, that source is *nature*. Not the nature we study and embalm in taxonomies (the Western view), nor the nature we point to and say, "There is proof of my Koan" (the Eastern view). Rather, I see our pursuit for blending taking us back to the *source*. (Chapter 13 addresses this issue in more detail.)

It is only in nature that the most persistent of open systems prevail. It is only in nature that we can view original holonomy uninterrupted and therefore more pure.

This creates an interesting perspective from which to view the contemporary dedication to conservation and preservation of natural systems. Is it possible that we are beginning to see beyond the encapsulated "nature" of zoos, parks and greenbelts to the central quality of wilderness?

It is possible that we need wilderness not for recreation or leisure . . . but for the guidance that will allow us to create a realistic, holonomic equilibrium and unity between the now distant and fragmented perspectives of technologized world views and the planet.

Cultural perspectives nurture stability. Although all cultures are forced to acknowledge change, cultures view themselves in static equilibrium models. It is the "outside" world that is supposed to change, to accommodate to each culture's closed-system perspective of its own basic structure.

In nature, there is no such perspective. Ecological systems have no closedness, nor do they defend political boundaries. Even the territorial tendencies of certain species are fluid and situational. Non-holonomic systems require boundaries, they require closedness. Holonomic systems are boundless. The possibility of planetary disaster and holocaust lies in tomorrow's movement toward non-holonomic world views. In non-holonomic vision, the planet—including its resources and inhabitants—is seen as grist for the maintenance of varying kinds of closedness. The religion, philosophy, political vision, and martial predisposition of many technological cultures act as the driving "truth" to be followed by the rest of the world. A comparable non-holonomic circumstance in nature would have zebras organizing and entrenching in order to make all giraffes, gnus, and baboons convert to zebra qualities. The zebra example instantly creates an image so ludicrous it can be dismissed quickly. Yet the zebra metaphor is basic practice in the churches, legislative bodies, schools, and war ministries of every nation on this planet.

Cultures and their non-holonomic qualities are reaching an impasse. Non-holonomic cultures are demonstrating the limitations of such world views. This dawning awareness is primarily linked to the snarled net of economics that has forced the reality of Marshal McLuhan's global village metaphor upon us. Linking this with the nearly instant communication of electronic media, we are all bound to the realization that closed systems are failing. Cultures are being forced toward open-system intercourse for their very survival. Technological ingenuity is being universally dispensed; Japanese, German, and Italian flags fly in conquest over large sectors of the American industrial machine.

American technocrats—like technocrats throughout the world—feverishly search their prior experience and history for guidance toward future fulfillment. It is becoming alarmingly clear that history affords us no guidance. Every aspect of closed-system problem solution has been tried. In the ten thousand years since *sapiens* planted the first seed, we have accumulated an awesome list of closed-system failures. Nationalism and disease were to be cured by war and medicine. Boundaries enclosing ideologies were fixed and maintained after the wars, with nationalism and the threat of war born again. Medicine cured diseases, contributed to over-population, and shifted the causes of disease from the swamp and the sewer to the dominion of hypertension and stress.

Non-holonomic vision has been expensive. Most of this expense has been due to interference with, lack of respect for, and destruction of natural holonomy. Plants appeared on this planet approximately 400 million years ago. They are still here. Fish preceded plants by about 100 million years. They too are still here. The roles of plants and fish on this planet followed holonomic patterning, holonomic harmony . It is appropriate to question whether *homo sapiens* will make it through fifty thousand years. It is provocative to reflect on whether or not *sapiens* will re-establish holonomic world view. If so, where will *sapiens* look for vision? Will Westerners keep looking to the East for the way? Will Easterners keep up their voracious Westernization? Will both look to the stars and see new worlds to be ravaged via an interplanetary exploitation of non-holonomic vision?

Even on my most cynical days, I cannot believe *sapiens* to be so fragile. Instead, I see a reunion with holonomy. People will look again to nature . . . look again to the earth.

I see it in Peter Matthiessen's words:

> When modern man has regained his respect for the earth, when science becomes a tool in the service of nature rather than a weapon to dominate it, then the lost Paradise . . . of all people on earth will come again, and all men will be 'in Dios,' People of God. [4]

PART II

Culture and Mind

A change of focus. Whereas Part One was an exploration of some theoretical and scientific postures regarding the mind, this next portion of the book will provide a new background. This new background will be expressed through the medium of people. . . . human perspectives as seen against the larger tapestry of culture.

Chapter 6, *Toward a Holonomic Culture*, is an essay on culture. This perspective is richly endowed by Charles Reich, an historian who wrote of cultural evolution in a somewhat ecological sense, although he did not use the term. As I interpret Reich, he paints a broad-brush portrait of the evolution of American technological society.

Chapter 7, *Holonomy and the Person*, extracts the person out of the fabric of culture and examines the individual personality. The perspectives of O.J. Harvey are central to this chapter. Harvey sees individual people and their personalities as exemplars of the elements of society to which they express allegiance. Harvey's work clearly provides detail as to how individuals may be both the sculptors and sculptures of the social ecology.

In Chapter 8, *Evolution: the Person and the Culture*, we explore the mechanisms of evolution, including a description of how the process creates options through which individuals express their roles in a culture. Social or cultural consciousness establishes human ecologies . . . yet the collective we know to be society is but an expression of an aggregate of individuals. In a sense, an individual may enter the evolutionary process at will . . . *but only as an individual.* Human beings have the capacity to *give themselves consciously* to evolution .

FEB. OREGON .51

6

TOWARD A HOLONOMIC CULTURE

According to some perspectives within the discipline of social psychology, individual human beings may as well not exist—at least not as a unit of study. Each person is seen as part of the family of humans. Many sociologists and social psychologists ignore the individual and focus their study on immediate families, tribes, cities, states, nations, and finally humankind as a whole. The individual is seen by these social scientists as being born into a family, the family is seen as an element of a culture, and a culture is viewed the result of a collective with agreed-upon closedness in thought. Thus, in a powerful sense, the individual becomes little more than an integer in the collective sum which describes the culture.

The process appears to go like this. Once cultures establish fixedness in their thought, they create coercive strategies that make the individual conform to the fixedness. These have many names: values, morals, attitudes, beliefs, and the like. All of these can and have been systematically studied. Hundreds, if not thousands, of scientific observers have done so. For example, most anthropologists historically have begun their study by first establishing the results of a group's closed-system thinking as expressed by the folkways, mores, and laws.

In view of this "collective" vision, individuals in any group are expected to conform to these agreed-upon consensus positions. They are generally—to a wide degree—punished if they do not conform, and rewarded if they do. In the most altruistic interpretation one can, at best, say that each individual is born into the safety and protection of like-minded people. At worst, one can see the newborn as a pawn of a group of single-minded despots who mold children into conforming elements of a fixed world view.

As harsh as the second of these postures sounds, it is likely that most cultures on the planet are predisposed to behave like this despotic example while explaining their role in altruistic terms. Nearly all of the institutions of a culture (particularly our own), portray themselves as altruistic; yet, when one looks carefully at them, it is clear that they all possess mechanisms of punishment violent enough to make nearly all relent and comply. The family in our culture has, in times past, been the initial communicator of the culture's expectation.

The point of all of this is that, with respect to the holonomic potential of most individuals, cultures are systems that to varying degrees systematically limit the realization of holonomy. It is the closed, non-holonomic portion of culture that creates passports, national boundaries, and wars. The non-holonomic is essentially the cultural citizen, while the holonomic is basically the natural human.

This is not to say that holonomic families cannot exist. They do . . . and they are characterized by an honoring of the individual members as their primary function. Their focus is on the development and celebration of the whole human. They are not devoid of concern for the external "control" tendencies but these are secondary to the commitment to nurture the natural and inherent qualities of humanness in each family member. Non-holonomic families keep the locus of control, the place where motivation toward behavior resides, outside the individual. They prevent the individual from having control over their own destiny. The holonomic family nurtures an interrelatedness of self-reliant individuals. The non-holonomic family requires dependence on external doctrine.

Closed-system non-holonomic thought dominated the very founding of the United States. A reaction against the crown and church of England, and the chance to own land in the New World, gave birth to an ecology of reactionism that formed a significant branching of history. It was an era in which eager conquerors took a virgin wilderness, exploited its resources and ultimately created the most awesome technocracy the world has yet known.

More than a decade ago, Charles Reich, in his epoch-making book, *The Greening of America*, called this first establishment of American closed-system thought "Consciousness I." In the colonies and early United States, it was represented by an anti-authoritarian backlash against the version of closed-system thought exemplified in the Old World. The fervor of this passionate abandonment of Old World contexts resulted in a system nearly as oppressive as the one journeyed from. Citizens of the New World grasped tenaciously on to the opposite sign of their Old World experience. They established a rigorous code of conformity that mirrored the wounds of righteousness from which they fled. Moreover, they extended the new morality to a totally new group of people . . . the native Indian.

In terms of the individual, there was little liberation. Following the exodus from Europe, the tendency to abandon the old prevailed in an almost paranoic way. With no *new* set of folkways, mores, and laws immediately available to the new settlers, they inadvertently gave birth to an extremely subtle new tyranny. Reich describes it this way:

> Facing a new and vast land, a new freedom, and seemingly limitless riches, its reality centered on the truth of individual effort. America would prosper if people proved energetic and hard-working. The crucial thing was to release the individual energy so long held back by rigid social customs and hierarchical forms. Each newly sovereign individual could be the source of his own achievement and fulfillment. One worked for oneself, not for society. But enough individual hard work made the wheels turn. Consciousness I focused on self, but it saw self in harsh and narrow terms, accepting much self-repression as the essential concomitant of effort, and allowing self to be cut off from the larger community of man, and from nature (defined as an enemy) as well. This uniquely American consciousness expressed the realities of the new nation. [1]

So while thoroughly convinced of new liberty, the American vision created a tyranny of the self as opposed to the tyranny of the group. Each person awash in this tyranny had full access to developing a more open context of thought. Yet few did so. It was found that the revolutionary who wins inevitably becomes more oppressive than those from whom liberty was sought.

The new American did have the freedom to escape the confinement of closed-system thought. Yet the Anglo traditions of repression and discipline died only in name. Thus, the new American created an alternative personal vision of closed-system thought . . . and immediately began a harsh term as judge, jury, and executioner of one's self.

As Reich and others have said in describing this context, the new American saw nature as the enemy. Nature—the earth, its trees, and its waters—were the enemy to be conquered in a long war to be waged in exploitation. So the individual, divested of the Old World's brand of closed-system thought, created another which was unique to the New World. This new brand of closed-system thought penned a virtual death warrant for the New World's wilderness. The Consciousness I humans for whom holonomy had been excluded by Anglo-Saxon heritage now locked holonomy out with their own souls.

From approximately 1620 to 1790, Americans institutionalized a form of reactionary ethic. Although it has been characterized as the Puritan ethic, it was clear that its overtones of individualism and self-reliance were being eroded subtly by the next great context of consciousness. In 1790, nine out of every ten Americans worked the land. Thirty years later, in 1820, the population reached nearly 10 million . . . and about 80 percent worked the land. Already the percentage of those who worked the land was dropping. Where did these people go?

The infectious consciousness of the industrial revolution began to permeate the New World. And with it came the second of Reich's Consciousness eras . . . Consciousness II. Whaling in the New World had expanded its reach across the

Pacific to Japan. In New York, a process by which salmon, lobster, and oysters could be canned in metal tins was invented. Lathes to tool steel had been fashioned, and President Monroe sealed the fate of Native American Indians with a proclamation that *hunting and gathering required far too much space. . .* far more than "civilization" could afford.

Machines were emerging . . . machines that would insure the introduction of systematic strictures to thinking required by organization. The new continent was populated by rugged individualists who were rapidly losing their individualism to support the growing industrial inclination in America. As the riches of the land produced surplus, the land was able to support the non-agrarian class. Machines were the forte of the non-agrarians. They used them to manufacture railroads, steam ships, and Elias Howe's new sewing machine.

With the birth of what Reich called Consciousness II, an ethic of technocracy emerged. Individualists who still needed to be judge, jury, and executioner found solace in the westward expansion. Those who felt a yearning for an external context of personal governance rallied around the dawning of corporate maternalism. The coordinator of commerce, the mother of manufacture . . . the *corporation* was born.

To be sure, the brand of closed-system thought that pervaded these times was perceived as being different from Mother England's and that of the other parent nations. The period of the Puritan Ethic had been instrumental in providing folkways, mores, and laws for this foundling culture—but had provided nothing to serve the aching desire for an external context of self governance. Legally or illegally, depending upon one's point of view, the bond to England had been severed. But the documents of law that had been invented to declare that independence would stand more than two hundred years before being understood by the people whom they serve. Understanding would come with the demonstration of the failures of the ethics of Consciousness II.

The law was not the moral force of the New World. *Industry* was. Profit was championed over justice. To gain profit in the arenas of commerce, organization had to be discovered. Moreover, it had to be implemented. It would take until the 1900s for this context to gain superiority, but, once established, it became so massive that the entire planet would have to reach its limits of consumption before it could topple. When it happens, its final death throes will be agonizing.

But what of thought and consciousness? The technocratic ethic is one with goals of growth, organization, and consumption. It requires profit to sustain these goals. It is an ethic based on methods of organizing responsibility, and a corrupt mode of judgment which demands conformity to the models of organization. In short, the technocratic ethic provided the solace for which the New World populace was yearning. It provided a substitute for the reactionary set of values, morals, and ethics imposed when the New World was established. The impact of running to the New World had been too severe, even for the rugged individualist who made the journey. It had created a view of self and individualist far more harsh than anyone could have expected. The technocratic ethic promised a sense of belonging, progress, and the kind of allegiance that would return the iconoclast to the community of humans.

Unfortunately, it turned out that the cost was great. The methodologies of the technocratic ethic were such that the very act of belonging shifted the basis of worth almost instantly from self to others. To reminisce a moment, make no mistake that those ensconced in the Puritan Ethic paid allegiance to the external, but the external was an idea—a model—they had selected. Being nonversed in alternatives, religiosity had provided the primary model for the Consciousness I citizen steeped in the Puritan Ethic. Surely religiosity is external, but—lacking a central coordinating agency such as a government or a history of tradition, as they had experienced in their former home lands—the

colonists of the New World chose a new closed system by which to define their individual virtue. In their previous home lands, religiosity had been administered by all facets of their societies; the citizen of the New World had no support base other than self to administer this ethic. Each person during the period of Consciousness I adhered to an individually-imposed version of an external ethic and external conscience. Now, with technocracy and the birth of Consciousness II, a new authority emerged . . . It was not the severe authority of the era of the Puritan Ethic, where all was black and white. The new ethic invoked authority in gray.

The gray zone between black and white has a metaphoric advantage. It can be what it describes itself as being. People yearning for externally-imposed standards and simultaneously yearning for profit and gain in consumption potential were more than willing to join a team. Teams were created in the context of organized manufacturing. But leisure time was not to be idle time. Sports were created that matched a teamwork and dependency ethic. Baseball was formalized by Abner Doubleday in 1839. Rutgers and Princeton played their first intercollegiate football game in 1869. Trying to fill the gap in team sports that existed between the baseball and football season, James Naismith nailed two fruit baskets to tall poles and invented basketball in 1891. Stanley Cup hockey began two years later. Meanwhile, the country had engaged in enough wars, both domestic and foreign, to insure that the consciousness of organization, authority, and flow charts of responsibility was preserved.

To the individual, this ecology became awesomely pervasive. Ethics were nurtured on these bases: belonging, fitting in, being part of a "team." In addition, I firmly believe that since all of these external contexts were rule-bound, the technocratic citizen began to establish self-worth in terms of playing by the rules. There was a continuous and insidious invitation to judge one's ethical legitimacy in terms of conformity to an external ethic. This was not a self sufficient type externalism-as was true in the era of Consciousness I-but, rather, a conformity to an external ethic of following the rules of corporate and commercial virtue.

At the base line, the one necessarily followed the other. The Puritan Ethic had nurtured a context in which nature was the enemy. The technocratic ethic labeled nature as resource/enemy and armed its citizenry with the tools of conquest. To sustain this conquest, the ethic had to be personalized. And strangely, all of the institutions of the culture—the government, the churches, the schools—formed themselves in the industrial model. Citizens, for the most part, were more than willing to personalize this context.

At the most personal level, a hedonism of consumption and acquisition was born. To produce and manufacture, the technocratic ethic demanded consumption. To insure that consumption, the team sports of buying, consuming, replacing, and expanding became the foci of the individual. The family became the team. The ethic of being self-sufficient was quickly replaced by group reliance; and a vision of self emerged in which self was defined by the degrees in which one was facile in externalizing. The individual was judged only in relation to others. Self itself became a resource for the external.

Closed-system thought for the citizen of the technocratic ethic was a complete and willing offering of one's self to the external rules of exploitation and continued consumption. For the citizen of the Puritan Ethic, closed-system thought was represented by a reflexive creation of an all-powerful, antithesis of prior experience. The new citizen in the Puritan Ethic was a mirror image of that which he or she had fled. In flight, these new citizens were owned totally by that tyranny which had spawned them. Without a whole array of cultural entities to enforce their reactionism, they created it within their domain of self. Thus, as rugged individualists, they became the antithetical pawns of the vision they hated. They were so absolutely non-reliant that it took only four generations of these people to transfer their pseudo-self-reliant posture to the technocratic ethic.

The technocracy called America had a continent so vast that its resource base seemed infinite. Whatever it needed it wrenched from the earth. Its citizenry moved dominantly from agriculture to manufacture, using the rest of the world as a marketplace for goods. Two devastating, yet corporately profitable, world wars had insured that marketplace. We devastated nations; we helped rebuild them and simultaneously nurtured expanding consumption at home by those "seeking the good life."

In our altruistic extension of trade, it became an alarming fact that those we had bombed out were rebuilding within the aura of our own advances in technology. New steel mills in Germany and Japan splashed envy across the corporate board tables of U.S. Steel, Carnegie, and others. These new mills were nearly a hundred years ahead of the museum pieces that produced the steel for Ford, General Motors, Chrysler, and others. New mills in Germany, Japan, Italy, and France began to feed the appetites of Volkswagen, Datsun, Toyota, Fiat, and Renault. Suddenly, our closed-system contexts began to open up.

But openness in closed systems must be represented by more than consumption. More consumption merely closes systems at a higher level. Consumption represents life for consumers, but death for producers. Another satisfied customer is one less that producers can depend upon next year. After all, who will buy a refrigerator every year . . . and how can the producers hope to reseduce year after year?

As pressure built up, the American dream began to represent the American nightmare. Seduction turned into rape. Through the medium of advertising or, better-labeled, commercial, promotional propaganda, we were led for a brief time to believe that we lacked grace, dignity, and even morality if we did not consume. We were tacitly asked to define our egos in terms of consumption. Our sexuality was questioned if we did not adorn ourselves with Thunderbirds, digital watches, and Brut . . . with Playtex, Chanel, and Clairol.

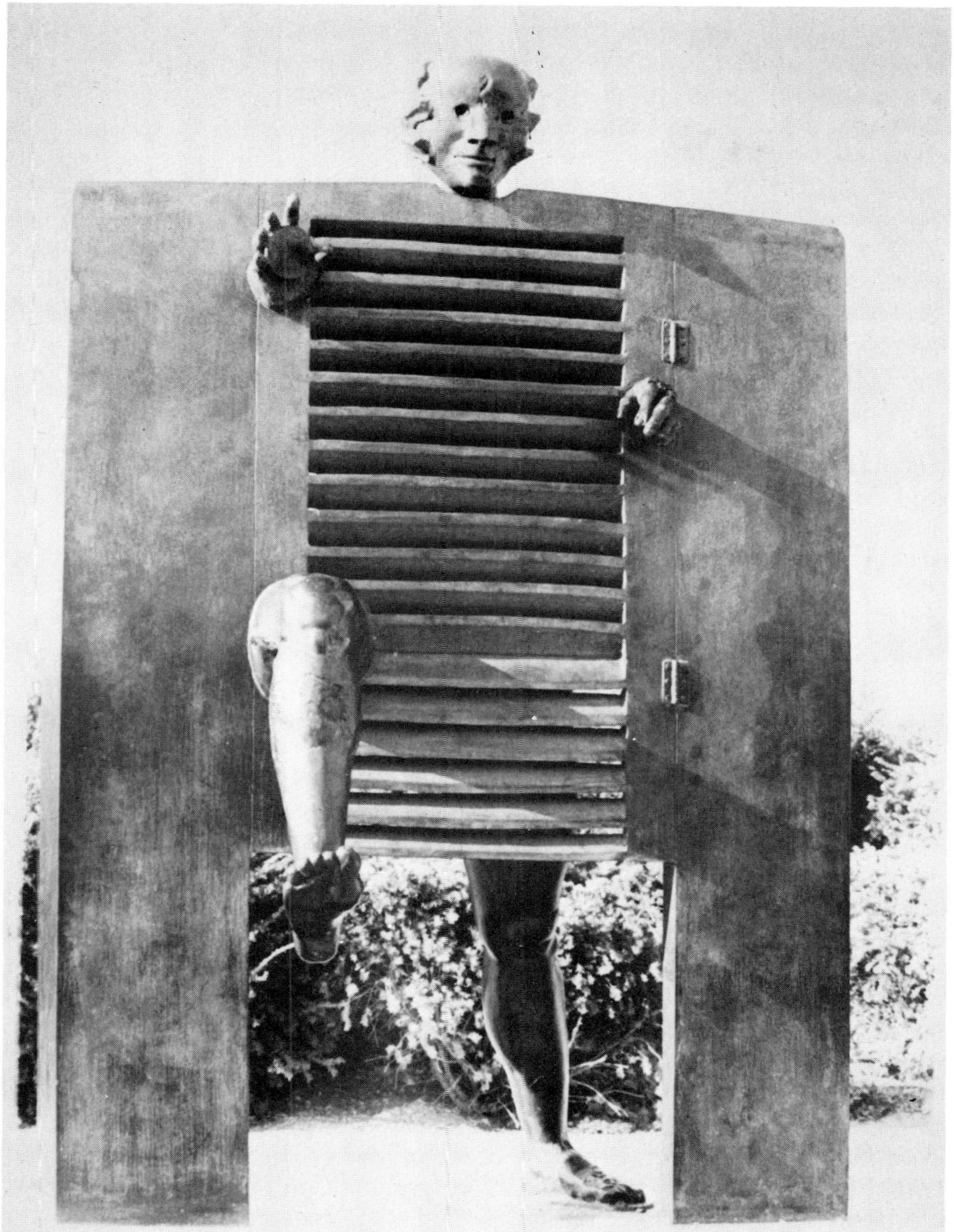

With all of this, television began subtly to live up to Marshal McLuhan's claim that media would create the global village. At first, his utterings seemed preposterous. But then, subtly and tacitly, the shift began. Tonight I will watch two hours of television news on a Japanese set and I will witness commercials in prime time for more Japanese, German, and Italian products than for products made solely in the United States. I will also see coverage of news that testifies to the interrelationships between nations. The technocrats thought it would stop at products . . . but it did not.

Tonight I will be exposed to cultural pluralism that none of my brainwash-oriented texts ever gave me in school. Governments will be shown tonight in those two hours to be vast aggregates of coercion and deception as they carry out the "noble" arts of diplomacy. Politics here at home will be seen as an extension of the deception of the pro-football quarterback, the relief pitcher, and the manipulative maneuverings of the foreward guard in basketball. The sports of teams have become the sports of deception. The acts of deception and manipulation have become the core of technocracy. The basis for such deception is the basis for the multinational corporations. All of it is called shrewd business, but it is basically and unrelentlessly . . . deception.

Yet there is emerging a cadre of explorers of the post-technocratic ethic. Confused, many seem dispossessed. At a loss to maintain blind allegiance to the technocratic contexts, they seem no more sure about what to do than did those who knew they must create a new ethic upon leaving the Old World for the new. It is these people who gave hope to Reich as he described what he called Consciousness III.

Consciousness III is seen to be comprised primarily of humans whose commitment to self is not based on the exploitation of others. They are not prone to exploit self either. Their visions of self-reliance are woven into a hopeful and supportive society. Yet this society is still emerging. It is one without the authoritarian adherence to doctrine found in Consciousness I. It is also devoid of allegiance to the manipulative, technocratic structures of Consciousness II. It is, in short, more holonomic.

Some of the visionaries of Consciousness III are seduced back into the social technologies of the past. Others are drawn to aspirations for new technologies by which to express selfness. This second group comes closer to representing themselves as integrated spokespersons for what Reich calls Consciousness III, and for what I envision as an intrinsic ethic.

Critics also have a tendency to look backwards. Christopher Lasch, in his bitter attack on the notion of an intrinsic ethic and the concepts embodied in Consciousness III, *The Culture of Narcissism: American Life in an Age of Diminishing Expectations*,[2] seems to miss the point while getting it. To me, Lasch is a nostalgist who searches for the past in a swampland of contradictions provided by the present. History has no capacity to predict the future, nor the capability of rendering the present objectively understandable. Interactions between technocracies is not a consciousness issue . . . it is a political issue. Lasch clearly confuses politics with consciousness. He sees our new search for selfness (via the ethic of the intrinsic) as an act of hedonistic, narcissistic politics. He seems not to recognize that politics is the issue. His judgments of the individual are based on his bias toward the group. Lasch fails to differentiate his political predisposition from consciousness which is an individual issue.

Nor should Lasch and his peers understand consciousness. Their role is that of spokesperson for the collective group and not the individual. *Again, consciousness is an individual issue.* And, as contradictory as it seems to most historians, *cultural evolution as well as natural evolution is born in the individual.* Later, it is exhibited in the group. This concept is expanded in chapters 8 and 9.

History, through the interpretations of historians, frequently seems to have assigned the role of evolution to the political actions of leaders—people whose actions have supposedly changed the world. Not so. It is far more likely that large masses of people, as individuals, create at once a tacit but understood change in consciousness. The "leader" only becomes spokesperson for the dynamics of the intrinsically-recognized plan for change. By assigning the intrinsic role to the leaders, historians seem victimized by charisma. Followers *choose* to follow . . . but their progress is seldom documented. Leaders provide historians with documents. The lifeblood of historians is the nature of documents, rather than consciousness.

In Zen Buddhism there is a saying, "When the student is ready, the teacher will appear." (anon.) It is antithetical to the technocratic ethic, and its philosophy of closed-system thought—thought owned by the external rules of enforced affiliation—that revolutions, or my preferred term, evolutions, be leaderless. The politically-biased are discomfited by leaderless change.

The point here is to champion an ethic of the intrinsic, not to attack colleagues and their work. It is not my desire to deny the historian's observation that followers sometimes choose blind, vacuous patterns of following. Yet, exaggeration often makes a point clearer. And it is my hope that this point is simple: *The emergence of the intrinsic ethic is born of those who would seek self as separate from the blind following of authority, separate from the blind belonging to technocratic game plans, and yet endowed with a commitment toward holism.*

Many, including Reich, mistakenly interpreted the reflexive anti-authoritarian response of reactionaries as a manifestation of the intrinsic. I would claim these are not intrinsic in their motivation at all. Many reactionaries are motivated by an immature passion for destruction. The intrinsic is a creator rather than a destroyer. It is possible that much of what infatuated Reich in his descriptions of Consciousness III was the burgeoning presence of this reflexive anti-authoritarianism. This anti-authoritarianism clearly was for the most part mindless activism rather than a broad-based commitment to intrinsicity.

The reactionary sixties were richly endowed with such reflexive separatism. The intrinsic person who has discovered intrinsic consciousness does not see life as a game of separatism, but rather as a recognition of unity.

Reich's critics cite his over-indulgence in symbolism such as bell-bottom pants and long hair. I am convinced that evolution of consciousness is accompanied by substance rather than symbol, and that many of Reich's critics were in fact those who were focused on symbols.

It is likely the intrinsic person will recognize the meaning of the subtitle of Lasch's book, *The Culture of Narcissism: American Life in an Age of Diminishing Expectations*. The planet is clearly signaling an age of diminishing expectations. But the diminishment is in the expectations humans place on the *planet*. The diminishing expectations that Lasch meant were cultural not planetary. He lamented the lowering of expectations toward conformity and the continued patterns of exploitation. We are experiencing a freedom from the historical traditions of conformity and from the gluttonous demands for the continuity of exploitation inherent in the technocratic ethic. Those who chastise the diminishment of the work ethic and the lessening of adherence to technocratic values are judging Consciousness III by the values of Consciousness II.

History is losing its non-fiction credence in the face of planetary survival. The intrinsic ethic is not a reflexive reaction. It is rather an expression of a trend toward natural unity. But the unity in question transcends corporate internationalism; it transcends parochial allegiance; it transcends politics and nationalism. The allegiance of the intrinsic ethic is growingly directed toward the planet . . . the cosmos. Technology is not about to be destroyed . . . it is merely being weaned . . . weaned of the blind servitude of technocrats who have ignored intentionally the inevitable evolution of their own consciousness.

Pluralism that goes beyond culture and into the domain of nature is being considered. We are beginning to realize that the division between culture and nature was an expression of chauvinism to begin with. Cultural pluralism is the first step toward resolution. It, more than any other effort, will create the possibility of a context of harmony among humans. Perhaps it is but another small step toward the realization that human harmony with non-human, living, and non-living systems is necessary to achieve in the re-establishing of holonomic unity.

Holonomic unity then is the death of divisiveness. We are exploring the world of non-divisiveness . . . it is surely a new world because human history is burdened with separatism as opposed to unification. The emergence of the intrinsic ethic is antithetical to narcissism. It is an ethic for which narcissism is cancerous. An intrinsic person cannot be narcissistic, for such a person is in reality a citizen of holonomic being. Such people belong to the cosmos. If there is a closedness in their minds, it is to the concept of closedness. The intrinsic is the threshold of universal openness . . . an openness that views all systems as open systems. Stability, and what may be called structure, will most certainly exist. There will be law, and there will be adherence to law. But there will also be an inherent realization that all culturally-created structures—whether folkways, mores, or laws—are as subject to holonomic principles as all other patterned energy in the universe.

The subtlety of evolution has shown that holonomic harmony alters itself when the patterns become too fixed. Humans who adopt the intrinsic ethic will seek stability and security, not in fixedness and control, but rather in openness and harmony. Thus stability and security will become themselves, evolutionary. Our concepts of achievement and growth will also be gauged by our commitment to holonomic evolution. Continued achievement and continued growth will become focused on harmony rather than exploitation. The acquisition and accumulation of grace will replace the acquisition and accumulation of material.

7

HOLONOMY AND THE PERSON

In the previous chapter, I reviewed the various cultural ecologies that have characterized American life to the present. The ecology of the Puritan Ethic of Reich's Consciousness I is characteristic of much of the early history of the United States. At that time, many individuals appear to have chosen to establish rigid, disciplined values representing a narrow range of options. Right and wrong were clear and were often specified in religious-moral doctrine. In the absence of elaborate institutions to enforce the New World's morality, each person psychologically ingested the responsibility and became an adhering spokesperson for the group. Individuals became judge, jury, and executioner of their own and other's morality. They were the rugged individualists of romantic history. Psychologically, they were actually quite the opposite of what could be termed individualists; their consciousness was an incarnation of an awesomely rigid and external group ethic.

With the advent of the industrial revolution, a major shift in collective consciousness took place. This shift to Consciousness II was created by persons who were clearly willing to lighten themselves of the responsibility of carrying a moral vicar within their psyche. They now had institutions, including corporations, to give the responsibility to. They made the vision of their personal worth become a vision of belonging. Self became a servant to the support of institutions. The technocratic ethic ingested the Puritan Ethic. Population growth and exploitative commerce demanded the externalization of the rules and those who would enforce them.

Near the end of Chapter 6, my interpretation of the ethic of Reich's Consciousness III was explored. This was the intrinsic ethic. Today, it is clear to some observers that the ecology of the intrinsic ethic is forming; but as of now it is far from complete. Even so, its presence is becoming tacitly understood. Reich explored a wide variety of symptoms and effects, but was cautious about causes. Disenchantment, disenfranchisement, and disillusion are often cited as characteristics of this ecology, and perhaps this is accurate.

The intrinsic ethic, however, will not emerge as a movement *away from* the Puritan and technocratic ethics. Instead, I believe this new ethic will emerge as a shift in basic consciousness. This new consciousness is neither a reaction against older world views, nor a statement of revolution. Rather, this new context of consciousness is a product of the evolving status of the individual. This is why the process is not one of politics, not one of revolution. It is growing clear that individuals are becoming more aware that the institutions we have created cannot guide us forward. It is more likely that social institutions have become so entrenched in the ethics of the external that it is hopeless to look to their reconfiguration as salvation. If anything, it is clear that group social and political reform is probably a mechanism of history, rather than of the future. It is individual humans who possess a quality of being which may lead us toward intrinsic consciousness. We will consider this evolution more deeply in a later chapter.

No one has studied the ways in which individuals sculpture their personality better than O.J. Harvey, an esteemed scholar at the University of Colorado. Harvey's work spans several decades; throughout this time, his work included study of the various ways that social institutions affect the establishment of specific personality configurations. Harvey's work suggests that the four systems of personality configuration described by himself and his colleagues have significant meaning. They are systems that clearly separate individuals' general and specific patterns of response to stress into four major categories. For the many who have applied the findings of this research in education, industry, government, and a bevy of helping professions, such a contribution is sufficient reward.

My primary commitment to Harvey's systems was related initially to the accuracy and grace with which use of these systems provided people with means for checking out *their own* personality types. My colleagues and I used these systems in working with thousands of educators in the late sixties and seventies. Teachers and administrators in public education are subjected to a unique array of stresses. The most poignant of these stresses may be their daily attempt to divide their own psyches in order to engage sincerely, compassionately, and effectively with between thirty and one hundred young people and peers.

The near impossibility of maintaining a sincere, caring approach creates the primary stress. Heaped upon this are the bureaucratic tendencies of the schools, evolving social patterns, and diminishing funding for education. Unionization has also contributed a separatist political quality to the whole enterprise.

Thus, we applied Harvey's systems primarily as self-descriptors for the educators to use in characterizing their own stress and coping patterns. Later, we began the process of inviting the people to intervene in their own stress processes. We were tentative at first, and became much bolder with time. Our boldness was born of the success so many had in shifting their response pattern from one system to another. In other words, once individuals knew their own patterns and knew the pattern of less stressful ways of coping, *they were able to change*.

Harvey's systems were not only descriptive, but *they became representative of an evolutionary sequence*. This was an evolutionary sequence entered willfully, intentionally, and by each individual.

For me, then, the most intriguing facet of Harvey's work is in how it provides a structure for understanding a possible evolution of personality. Personality, as I use it here, is essentially the motivational/behavioral expression of the consciousness of an individual. Although Harvey's original objective was not in discovering a means for understanding the evolution of personality, it is clear that his categories of core personality lend themselves to an evolutionary interpretation. This is particularly true when Harvey's systems are superimposed over the evolutionary patterns of consciousness discussed in the previous chapter. But first, let's consider Harvey's systems:

SYSTEM 1

These individuals show greater dependence (relative to other systems) on external authority, relying on external sources for their standards of conduct and criteria for evaluation. These external sources are generally extra-personal forces such as God, norms of society, institutionalized authority, tradition, etc. System 1 representatives prefer high structured situations and display an intolerance for ambiguity. A poor delineation between means and ends is accompanied by a strong commitment to '*the* right way' to do a given task, and thus the search for multiple or alternative paths and the willingness to consider new information are highly limited if not prevented completely. This situation results in stereotypy in approaching problems, insensitivity and resistance to environmental inputs which are not congruent with the existing cognitive organization, and low ability to change set. In addition these individuals tend to show ritualistic adherence to rules without understanding, high religiosity, high absolutism, high evaluativeness, high identification with social roles and status positions, high conventionality, and high ethnocentrism.

SYSTEM 2

This style of functioning is characterized by negativism and an anti-rule, anti-authority orientation. The cognitive structure is somewhat more differentiated than in System 1 . . . however the cognitive organization remains poorly integrated and thinking still tends to be fairly compartmentalized, as indicated by the inability of the System 2 individuals to envision the implications and possible effects of their rejection of some aspect of their environment on other aspects. Perhaps because of the ambiguity, vacillation, and inconsistency that they perceive in their environment, they associate unstructuredness with distrust, loss of security, fear of rejection, and/or a feeling of loss of control over their situation. Thus the need for structure and intolerance of ambiguity remain high but are likely to be manifested in suspiciousness and avoidance of commitment. Individuals from System 2, more than any other system, are in a psychological void, rebelling against structure and authority on the one hand and rendered fearful and anxious by the absence of authority guidelines on the other. The rebelliousness of the System 2 individual appears more manifest while the seeking of structure and the attempt to find security in stable authority is more latent. Thus outwardly the representatives of this system tend to display negative valence toward the same referents that are of high positive relevance to representatives of System 1; it is important to note, however, that both use these same external sources as points of reference . . .

SYSTEM 3

This mode of functioning . . . is characterized by a desire to be liked and by attempts to establish and maintain relationships that foster mutual dependency and allow for manipulation of others. In fact, System 3 representatives have come to rely upon dependency and manipulation of others as their primary technique of controlling their environment. This type of interpersonal experience results in an awareness of self as a causal agent, sometimes an exaggerated one . . . System 3 individuals are much less categorical in their evaluations and tend to base their decisions on the implications of effects they will have for themselves and/or other people. Because of their ability to exercise control over others, representatives of this system are less deferential toward authority than representatives of System 1, less negative than individuals of System 2, and in general less concerned with extra-personal forces and institutionalized authority. They are, however, very concerned with attitudes of peers, social acceptance, social responsibility, and the standards of behavior prescribed by their particular reference group. Since they do not develop clearly delineated personal standards, they are in constant need of feedback from significant people in their environment in order to regulate their behavior and attain the acceptance and mutual dependency that they need. System 3 representatives manifest the need both to be dependent on others and to have others dependent on them. Their dependency apparently is directed toward individuals of power and status while those whom they would have dependent upon them appear to be persons low in status, power and expertise possibly because such persons would be easier to manipulate under the guise of helpfulness . . . Fearful of facing a situation alone, where success would depend on individual performance and/or personally derived criteria, System 3 individuals are extremely vulnerable to the threat of rejection, social isolation and other social conditions that might prevent the existence or use of dependency relationships.

SYSTEM 4

This style of functioning, the most abstract of the four systems, is characterized by high task orientation, information seeking, exploratory behavior, risk taking, independence without negativism, internal standards of conduct, personally derived criteria of evaluation, and relativism in thought and action. The conceptual structure is more highly differentiated and integrated than the other systems. These individuals are able to consider a given concept domain from many points of view, evaluate the concept with regard to several dimensions and to see multiple relationships both among the several aspects which they are able to articulate within a given concept and between that concept and other elements of their cognitive organization. These individuals are open to new information and capable of integrating such information into their existing cognitive organization, making appropriate modifications in that organization if necessary . . . In general their ideas and attitudes appear to have been derived pragmatically from direct experience of environmental feedback and are not oriented toward adhering to externally defined 'truths' or conforming to inviolable social norms. They, more than representatives of any other system, work for intrinsic rather than extrinsic rewards . . . Unlike individuals of the other systems who associate unstructuredness with uncertainty, insecurity, fear of reprisal, fear of rejection, etc., representatives of System 4 interpret these conditions as indications of trust and respect and they welcome the opportunity to exercise their independence and behave in accordance with their own inclinations. Thus System 4 individuals display a low need for structure, a relatively high tolerance for ambiguity, an ability to differentiate between means and ends, an ability to articulate several ways of attaining the same goal, a capacity to 'act as if,' a high ability to change set and a tendency to avoid stereotypy in solving problems.[1]

Harvey's work has been exhaustive in validating and cross-validating these systems of personality. Central to this work is the consistency with which a person remains true to type when stress occurs. Stress . . . that great unifier . . . that instantly perceived crisis maker . . . that crushing fear of losing control that pushes us back to our own primal self. It is when stress occurs that these personality systems are most specific. Strangely, persons most consistent with patterns of Systems 1, 2, and 3 are nearly always under stress.

The primary explanation for this is that the persons in Systems 1, 2, and 3 are highly external in terms of their personal focus. System 1 people externalize the source of their being . . . their God, their guru, their knowledge. They set it on an altar of self-righteousness. They are believers in *the Truth*. More than just believers, they *know* they are right.

System 2 personalities generally are reflexively against whatever anyone else believes in. Most often, they are against System 1 types because System 1 personalities generally represent the ethical status quo of families and nations. The System 2 person is the willing pawn of external belief. These people have an extrinsic tendency to react against, whereas the System 1 person preserves the prevailing unifying "truth."

System 3 personalities externalize in that they are willing members of larger groups. They judge their worth on the basis of the ways the group perceives them. They are intent upon manipulating or being manipulated by the collective course of life. They believe power is generally outside them, and their eternal search seems bent on wresting that power from the external or being actively victimized by it.

System 4 people represent a departure from the above in that their primary referent of control is internal. They are committed to the idea that they personally are responsible for that which they believe and do. They often have beliefs as strong as System 1 people do or might, but rather than being privy to "the truth," they see instead their allegiance to the belief as an individual choice. As a result, they are far more compassionate and flexible in their judgments of others. System 4 people might well become engaged in revolutionary intervention or work faithfully in an innocuous bureaucratic career, but, again, it is clear to them that *they have chosen their involvement*. In short, System 4 people are more bonded to open-system formats of being. Their lives are described by an intimacy and comfort with open-system approaches. Any certainty and closedness, while sincere, is always flexible. They trend more toward holonomy.

What I see as I look at Harvey's systems is a powerful model of individual choice-making, perception of control, and psychological focus that is remarkably compatible with Reich's stages of consciousness. Reich clearly sees the stages of consciousness as evolutionary. But can this apply to personality?

Harvey is adamant about his systems *not* representing stage theory. That is, he does not see these as sequential stages of development in the sense that they represent an acquisition of specified skills. For example, Jean Piaget, a Swiss psychologist of profound influence, developed a stage theory of intellectual development. Piaget postulated a sequential development of skills of reductive logic. As a result of development, one "matured" through stages that resulted in higher capacities to think in more specified ways.

Harvey's systems do not work that way. In fact, the difference between the personality systems and stages is that the systems move from specificity to more generalizable and *inclusive* ways of being. Erich Jantsch, in his book, *The Self Organizing Universe*, clearly distinguishes between evolution and development:

> The evolution of the universe is the history of an unfolding . . . Unfolding is not the same as building up. The latter emphasizes structure and describes the emergence of hierarchical levels by joining systems 'from the bottom up.' Unfolding, in contrast, implies the interweaving of processes which lead simultaneously to phenomena of structuration at different hierarchical levels.[2]

It is vital for me to emphasize that evolution is different from development. Since it is an open-system process, any and *all* criteria we use to designate its flow are *simultaneous throughout all of its progress*. The individual is the primary element of evolution and the ecology that the individual experiences reflects the consciousness that dominates experience. But unlike the developmentalists' approaches, there is no external hierarchy. Individuals, within their own personalities, guide the process, unfolding links backwards and forwards. Thus, the evolution of personality becomes a process of interweaving the individual back to origins and forward to an eventual model of culture.

Recall my working definition of personality—the motivational/behavioral expression of the consciousness of the individual. As I see it, each person possesses certain motivations which are strongly influenced by the overall cultural ecology within which he or she is born and reared. The smaller the social unit, the more complete the exposure to the specifics of the expected conformity patterns. In other words, the individual in the hunting and gathering family has a very clear picture of the expectations that that cultural unit has of him or her. As the cultural unit increases in size, there are more mediaries and more confusion about expectations. Also, as numbers increase, there is a shift from a focus on motivation to one on monitoring of behavior.

The result is that in a small-number culture (family, clan, tribe), there is more direct, shared awareness of each person's expression of his or her true motivations. This is perhaps why tribal structure is seldom rigid in its organization and formality. Although there are chiefs, warriors, and shamans, people usually are chosen for their motivational linkage to the role, rather than as a kind of "promotion" born of a flow chart. A culture like ours in contemporary America requires such an array of superficial involvement that family members can seldom be concerned with motivation. Managing behavior is about all that can be hoped for. Quite often members of complex cultures such as ours are content adopting behaviors that are in no way central to their motivation.

Yet in both cases, consciousness affects motivation *and* behavior. Specifically, this means that if a person experiences a shift in consciousness, it is likely that new behaviors will emerge. Often, in the process of "trying on" a new kind of consciousness, a person will role play the behavioral expression of that consciousness.

The martial arts offer an example of this. In the martial art of Aikido, there is no attack form. It is a martial art devoted entirely to the "blending" with incoming energy. This means that if one is attacked, the aikidoist does not counter attack but, rather, intentionally blends his or her behavior (movement) in order to achieve the greatest harmony with the attacker. This blending is so antithetical to the concept of attack and defense that attackers are usually bewildered by the blending gesture and are rendered helpless.

While it is clear that Aikido involves behavior—that is, movement—it also involves motivation. The beginning aikidoist can learn the technique easily. Beginning students can master movement and style and do so without the slightest change in motivation. Their motivation stays locked into the concept that they are not blending with the attacker, but, rather, they are deceiving and tricking an attacker.

As the beginner progresses and mastery is approached, the behavioral "role" begins to affect the motivation. Or, in other words, the "body" begins to convince the "mind" of the compatibility with the philosophic overtones of this new consciousness. When the mind listens, blending becomes a part of the motivational base. At that point, one passes from awareness to assimilation.

It is this kind of student that progresses to the higher orders of achievement and harmony. For those who do not make this mind-body merger, the martial art is usually dropped.

This phenomenon of role playing or "trying on" of a new consciousness is a primary avenue for evolution. But, as anyone who has attended any "new age" type of conference or celebration knows, little need be invested to adorn the trappings of new kinds of consciousness. The charlatan of consciousness has afforded critics like Christopher Lasch and Peter Marin far too much grist for their denouncements and denials of such possibilities. A more hopeful scholar, Theodore Roszak, in *Person/Planet* writes:

> I call this insistent quest for personal growth and spiritual fulfillment 'positive'—but I do so advisedly, in full recognition of the mindless and meretricious nonsense one can find mixed throughout this burgeoning range of religious and therapeutic experimentation, as well as in the funky chic merchandise and life styles that seek to present themselves as 'simple living.' But I find myself far more concerned by the indiscriminate hostility that has been voiced against even the finest of these efforts in academic and intellectual circles by critics (often scandalously misinformed critics) who seem determined to see nothing new under the sun that is not folly and fraudulence.[3]

What I am saying is quite simple. Evolution in consciousness can be approached both through mind and body, through motivation and behavior. Because of this, the individual clearly can take a role in affecting his or her own consciousness. Individuals can affect consciousness with behavior and then monitor the degree of blending of motivation and behavior. Or, conversely, individuals may detect within themselves a continued dissonance between motivation and behavior. Both the blending and the dissonance are positive feedback. In one case, the result is a transformation in consciousness; in the other, a testing and eventual rejection of a new consciousness mode.

The key issue in all of this is my view that, regardless of one's personality system or cultural ecology, there is a growing realization of *personal* choice. People can choose the model of person they want to be. Most people at this moment have already chosen the personality they represent by retaining and reinforcing the qualities of the social ecology that spawned and nurtured them. But, unlike the claims of many, this personality is not fixed.

Zick Rubin, writing in *Psychology Today*, addresses the question, "does personality change after 20?"[4] His conclusions are guardedly open. Rubin quotes researchers who claim we are a blend of opposing forces. One quality of our personality strives for stability and continuity; the other seeks to be more than we are. Cultural influences historically have tended toward the non-holonomic. Fixedness and closedness are the most sought after qualities of being. Yet against the dynamic cacophony of change, it is becoming more evident that change is the only continuity . . . the only stability in contemporary times.

Humans are also realizing that fixedness is not the way to greet the realities of change. Only rocks and pebbles can depend upon gravity for so long in a mountain torrent. Sooner or later we are swept up and the patterns of change carry us along, whether we are prepared or not. In the past, preparedness called for more fixedness. Today the call is different. The call is for more flexibility, for greater capability to guide the change, for being enobled with the capacity to go with grace and in harmony with the dynamic shifts within human ecologies.

But does holonomy hold only the promise of "giving in" or "giving up," to use the rhetoric of the old paradigms? Not at all. Open-system ways of being do *not* strip people of their will or motivation. What they do instead is make people more realistic about the roles they play in what they pay attention to. Cultures have for centuries made their chosen formulae for stability and fixedness a required course in the lives of all their members. Holonomic consciousness is making it clear that all those required courses are in fact electives. People with personal histories marked by closed-systems experiences recognize their own exercise of choice. People who trend toward holonomic consciousness are coming to realize that they are far more free than they once believed. Moreover, they see themselves as the sculptors of their own stability and change, against a background of cultural stability and change. The two are related, but not equal.

When individuals release themselves from the effects of their own personal histories, it is likely that they move from culturally-induced closed systems to more nature-based open systems of thought. This move inevitably will be accompanied by a more open-system consciousness . . . by a movement toward the holonomic. Each shard of the liberation of one's psyche, one's motivation, and one's behaviors is a step closer to the mind of our mother. Each venture into exercising open-system being is a step closer to planetary and cosmic consciousness.

8

EVOLUTION: THE PERSON AND THE CULTURE

Charles Darwin was not the first, but he was surely the most popular spokesperson for organic evolution. He formulated a series of processes that serve to describe organic evolution. Among these processes were natural selection, natural variation, and mutation. The essence of each of these can be condensed into a sentence or two.

Natural selection, often callously called the "survival of the fittest," deals with the amazing number of interactions each organism has with others. Inherent in this interaction is the necessity for nutrition, space, and reproduction. Some live and some die. The characteristics of the ones who live are passed on to their offspring.

Second on this evolutionary list is natural variation. Natural variation has to do with the genetic differences between organisms. Creatures that reproduce successfully have high genetic identity between parents and offspring. The genetic makeup, however, is never identical. For example, although all members of your family share high genetic similarity, there are remarkable differences between each of you. Furthermore, each generation tends to experience a "drift" in certain nuances of structure. If these drifting genes produce advantageous results, then their continuity in the natural selection process will be higher.

The last of the evolutionary factors to be discussed is mutation. There is an increasing realization that there is a strong possibility for genetic codes to be affected by outside sources. Cosmic rays, radioactive isotopes, and the presence of "meddling" chemicals in diet might well create changes in genetic makeup that may be advantageous or not. Inherent in the process is the possibility that the mutation would be destructive and that continuity would be interrupted.

In nature, it is impossible to separate genetic processes completely, although working definitions exist. Each process is omnipresent, and all are part of the dance of life. When energy patterns stabilizing to create the harmonics of the genetic code are passed on to a mate, they develop a resonant harmony with the mate's patterns. The patterns blend and unite and begin the process of wedding their harmonies into new life. The newborn becomes a virtual symphony of holonomic energy. Its own musical score is the resonant macronic pattern conducted by its parents. But, like all offspring, its own music—slightly different from either parent—guides its own destiny. It lies within, waiting to be carried on at such time that new life, new harmony, is to be created.

All the while, organisms act, react, and interact in larger patterns. Some interact by eating others. As a result of this interaction, the "music" of some is often short-term, and replayed in the digestive tract of others. The patterns of energy configured in the processes of conception, gestation, birth, and growth in one creature suddenly begin to break apart and reharmonize with other contexts. At the end of pattern for one creature, the extension of pattern goes on. Death becomes the playing of new music . . . music of continuity.

Evolution in cultural ecologies and individual personalities proceeds in remarkably similar processes. Humans have tried since Darwin's time to link the processes first proposed in his theories. One early effort to apply Darwinian principles to social contexts was termed "social Darwinism." Social Darwinism was exemplified by the imposition of the concept "survival of the fittest" to human affairs. Unfortunately, "fittest" was seen at the time to be the last survivor in a sequence of acts of aggression. This spawned a long series of efforts to define "fittest" in terms of one particular set of folkways, mores, and laws, and to enforce these views with military, economic, and philosophical aggression.

Using a holonomic, rather than reductive process, it is clear that aggression and competition are minor elements in evolution. Unfortunately, we often have been persuaded to focus on the more violent aspects of the progress of natural systems. All of us have been harangued about territoriality, extinction, competition for food, and struggle for survival. Even our colloquialisms emphasize this "dog-eat-dog world," exclaiming, "it's a jungle out there," and declaring, "I'd rather fight than switch." Interpretations through history are not much better than those through the social sciences: we have all been inundated with emphasis on wars, revolutions, annexations, and insurrections in our study of the sequence of human affairs.

If one uses the metaphor of "pattern," then—with regard to its harmony—when one pattern disintegrates, it becomes part of a larger pattern. The creation of this larger pattern is as much a process of cooperation as anything else. Evolution becomes the process by which pattern changes through the continuity of life. More will be said about this in the final section of this book. In that section, I hope to clarify the apparent callousness of historical definitions which use competition in their efforts to explain evolution.

Holonomic approaches do provide a different perspective. Just as all the processes of evolution work simultaneously, so too do all of the elements of the holonomic process exist simultaneously.

At this point, I would like to present an evolutionary portrait of personality types. I will use O.J. Harvey's work as a basis, but I will modify freely and extend his systems to a model I believe represents tendencies not included in his system. Simultaneously, I will borrow from Reich, and will also modify some of his descriptions and conclusions. Figure V is the first aspect of this evolutionary model.

Figure V.

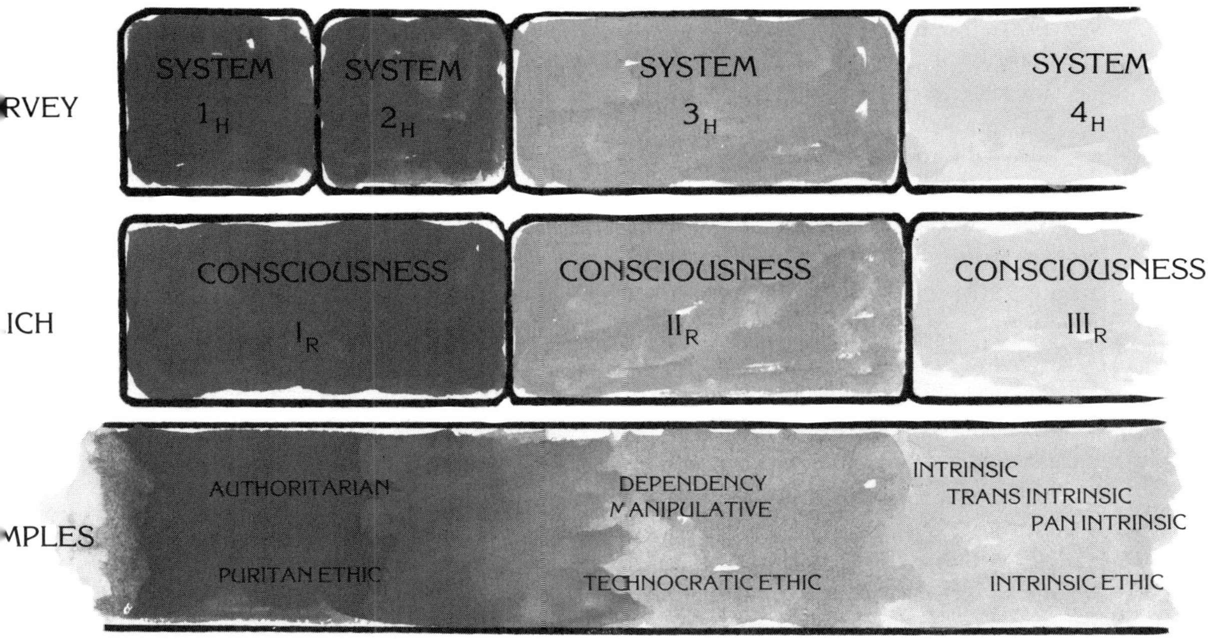

There is a remarkable blending of the ideas of Reich and Harvey, although their approaches are remarkably different. Harvey, a systematic researcher, has devoted a lifetime to the determination of personality types in individuals. Reich, a historian, looked at tendencies in the historical development of the United States that reflected shifts in consciousness. There is such similarity that I am surprised they were not colleagues or collaborators at some time.

Harvey's System 1_H and System 2_H—the authoritarian and anti-authoritarian types —were the typecast citizens of Reich's Consciousness I_R. The ecology of the Puritan Ethic was born of a dissent against authority. In all ways, the New World people represented both Systems 1_H and 2_H. Their anti-authoritarian tendencies (2_H) marked a reactionary departure from the old-order politics, religion, and economic base. However, once ensconced in the New World, their tendencies became quite authoritarian (1_H). Though the authority was not yet institutionalized, in the sense of extensive bureaucracies, the convictions of the citizenry quickly institutionalized it in the self. Thus, the rugged individualist was born. There was little trust in the integrity of this individualism, as is attested to by the early establishment of Harvard College, training ground for a new ministry. The rugged individualist, in effect, became the franchised version of prevailing external morality, and the populace lost no time in creating the enforcing marshals of this external vision.

As surely as Harvey's Systems 1_H and 2_H were created by and in turn contributed to the creation of Consciousness I_R, Harvey's System 3_H person was the fodder of the corporate revolution of Reich's Consciousness II_R. When America began to implement the technocratic ethic, there was a strong lessening of support for the rugged individualist in the economic sector. The Puritan Ethic or Consciousness I_R, did not die. It merely retreated to the homes, schools, and churches. There was little patience at work, however, for the individual. It was now team time, and people were expected to "work together to get the job done." Later, in Consciousness II_R, there emerged tendencies to create organization and formalized interdependence in the homes, schools, and churches as well.

Team sports, church socials, pot luck dinners, city government, girl and boy scouts, all paved the way for the flow-chart mentality that would create an America where worship, leisure, and love-making would fit into a schedule. In the presence of such an ecology of organization and belonging to the team, Harvey's System 3_H people both molded and were molded by the technocracy. Several major foreign wars created enough national stress to remind each citizen that survival depended upon their willing acceptance of their role as a gear in the large machine of corporate productivity.

Because our culture is currently immersed in the disintegrating domain of Consciousness II_R and the majority of those individuals who are currently "making it" fall into Harvey's System 3_H, the rest of this is primarily a fragmentary description of emerging tendencies. Both the qualities and descriptions of Harvey's System 4_H and Reich's Consciousness III_R people are rich in their use of the word, "autonomy." Autonomy is the most vital quality of the emerging individual; and a respect for autonomy is the most significant variable in the cultural ecology.

The transformation represented by this emergence may well be the most important in the history of humankind. No one has documented the tendencies characterizing this transformation better than Marilyn Ferguson in her visionary book, *The Aquarian Conspiracy.*[1] Although embryonic, this transformation represents a flowering of the possibility for a higher mutuality between the individual and the culture, one built upon respect for the autonomous self. Until recently, the autonomous person was victimized by the near-paranoid, group compulsion of Consciousness II_R. As such, human affiliation, so necessary for culture to function, has been the result of an obligated requirement to belong. The outcome has been that affiliation has been superficial and vacuous for many. More frightening than vacuous or superficial affiliation, was the thought of *not* being affiliated . . . the thought of being alone. Being alone in Consciousness II_R or as a System 3_H person meant mandatory loneliness—a condition worse than death—for these compulsive affiliators. The rehonoring, however, of autonomy seems destined to offer a remarkable alternative.

My analysis and synthesis of both Harvey's and Reich's work has resulted in the development of an evolutionary view consistent with the holonomic perspective. In evolutionary terms, much of Harvey's work focuses on the individual and the response capability possessed by individuals. His focus is on the organism. Reich focused on the ecology, the background system against which the individual is born, lives, and dies. My work in exploring the evolution of consciousness and in trying to relate culture and nature using models that unify in a holonomic way has resulted in the following conclusions:

Drawing upon the metaphors of harmony within change, it seems clear to me that evolution progresses out of a dynamic kind of reintegration of harmony. Disharmony makes substances relinquish their form (disintegrate) and become resources for reintegration in more harmonious settings.

Erich Jantsch, in *The Self-Organizing Universe,*[2] argues convincingly that *evolution is a self-organizing process.* Jantsch sees equilibrium as counterproductive to evolution. He has introduced the term *autopoiesis*. It means *self-renewal*. Jantsch's idea of non-equilibrium is consistent with what I refer to as "harmony." Harmony is *not* equilibrium, nor is it *non*-equilibrium. Instead, I use harmony to represent that dynamic, interactive, ever-changing process that results when discord is brought into blending dissonant elements into a functioning unity. Like Jantsch, I believe in a concept of self-regulation.

Jantsch and Ferguson in their separate works cite the theory of dissipative structures as being descriptive of the processes involved in this kind of evolution. Basically this theory is rooted in the concepts of holonomy that I attempt to describe. Since holonomic structures form because of tendencies toward harmony, so too do they re-form when the conditions change. The theory of dissipative structures is a systematic look at the transformations involved in the harmonic shifts. Illya Prigogine is most widely quoted as the spokesperson for these ideas.[3] In my mind the theory of dissipative structures is a special case of holonomy, just as holograms are a special case of holonomy.

Evolution, always starting with the individual, involves both organisms and ecologies (interrelating systems). In a human sense, we could say that evolution involves both individuals and cultures. Both are simultaneously and interrelatedly vital. Minute changes in the individual eventually result in subtle shifts in the ecology.

In holonomic terms, there is a mutuality to the harmony. But as surely as there is a mutuality, the music, the dance, changes. Continuity is based upon continually re-establishing harmony.

Historically, humans have externalized their focus as tendencies for evolution occurred. Generations of individuals gave their intrinsic freedom freely to cultural ecologies. The external truths or technologies of a people quickly began to own them. It is no wonder that when change in individuals became recognized as an option, the individual asked an external *Truth* for permission to change (Consciousness I_R). Or, in Consciousness II_R, individuals asked the technological system for permission to change.

The outcome of this compulsive externalization has been the tendency in individuals who demand change to believe that they must destroy the ecology or the system. These persons come to the gnawing realization that they are hollow humans. They have given their souls to the external. The ecology owns them.

These are Harvey's Systems 1_H, 2_H, and 3_H. These are the citizens of Consciousness I_R and II_R. This psychological externalization of their pattern of being leaves them vulnerable to the whims of the ecology . . . of the culture. These people, the Authoritarians and the Dependency/Manipulatives (See Figure V, page 97), relinquish the criteria for their being to outside forces. The Authoritarian, the true believer, appears on this earth only to represent a higher authority. They have no life, but the blind obedience to that external, designated belief that owns them. To the Authoritarian, life is a battle . . . *to be won*!

The Dependency/Manipulative person requires a virtual meadowland of other people whose psyches they can harvest. Their whole being is measured in terms of who they manipulate or are being manipulated by. Their commitment is to pawnmanship. They move about the board of life, knowing they are the weakest piece on the board, but hoping someday to take a knight or a castle. Some yearn only to be lost from the board early; others hope someday to move into an opponent's back row and find all the power that had been formerly used against them. To the Dependency/Manipulative, life is definitely a game.

In progressing, we come to Harvey's System 4_H person and Reich's ecology of Consciousness III_R. Autonomy . . . intrinsicity. This is a person who develops the capability for harmony with the external, but does not void self in the process. These persons recognize tendencies in the ecology that nurture evolution, those qualities in the larger contexts that move toward openness. Because they themselves experience open-system thought, they harmonize with elements of the cultural ecology which have tendencies toward more liberating contexts.

For me, the intrinsic state is a transitional state (See Figure V, page 97). It represents the position of many in our current culture. In some ways, I believe it represents a time when most begin the process of internalizing the locus of control. They begin to withdraw the power they had given to the external. It isn't instantaneous, nor is it smoothly linear; it happens in bursts. Often, with those who explore this mode of consciousness, there are preferred places where experiences are intrinsic.

With some, it may start with cutting down T.V. time . . . reliance on the tube is withdrawn. Others may stop letting co-workers annoy them. Still others may refuse to vote if they dislike all candidates. Once the process starts, one can accept greater responsibility for beliefs and develop a world view which begins to extend to all other humans.

This, in my terminology, marks the entrance into the Trans-intrinsic. Contrary to the Intrinsic, who is in transition, the Trans-intrinsic person is firmly participating in Consciousness III_R and shows clear qualifications for Harvey's System 4_H. The Trans-intrinsic person shows intrinsicity in *all* cultural involvements. Alone, with family, with colleagues, and in the world community . . . in all settings, the Trans-intrinsic is able to maintain an internality of responsibility.

The Trans-intrinsic person is all of the things spoken of in Harvey's description of a System 4 personality. The distinction is that Trans-intrinsic people have chosen to extend their life experience and personal history willfully into the lives of others. Their roles are those of teacher, manager, parent, administrator, and healer. They take action on life in the presence of others. But unlike the System 1_H Authoritarians or the System 2_H Anti-Authoritarians, they do not possess the *TRUTH*! Instead, they possess a sense of holonomic grace. If they are teachers, their students will witness the teachers' interpretation of grace. Their students will be encouraged to emulate the skills of that grace; but when they are judged, they will be honored for what they have done to personalize what they have been taught.

Trans-intrinsic people are celebrators of self, and of a growth-oriented open-system integrity. They know and honor the pervasiveness of holonomic change. Again, if teachers, their teaching role is dedicated to preparing students for open-system evolution rather than closed-system tenacity. These same qualities apply to management, administration, and healing.

Make no mistake, the Intrinsic and Trans-intrinsic persons feel emotion, pain, and non-agreement. But they do so from the base of knowing that they are choosing to allow themselves to feel that way. The Intrinsic types do not blame the ecology for happening *to* them. As such, the external certainly established a background against which life can be lived; but it is neither a stone tablet of law nor a gameboard of deception.

In a sense, the Intrinsic is a person with a growing sense of comfort that accompanies becoming open enough to know that closed systems are usually self-made. Trans-intrinsic persons have that sense established and are confident enough to approach others without burdening them with the inhibitions of closedness. Being open and establishing open-system thought with others is awesomely effective in establishing an honest closeness. The Trans-intrinsic person, by example, nurtures movement away from externalization in others. In doing so, the individual begins to affect the ecology, the personality system begins to affect the consciousness.

The next context in our scheme is the Pan-intrinsic. Whereas the Trans-intrinsic person has established intrinsicity with culture and its people, the Pan-intrinsic person has extended himself or herself to natural systems as well. The Trans-intrinsic person engages with people in comfortable, open-system ways which must also be achieved by the Pan-intrinsic person. It is in the recognition that to achieve a truly holonomic view that nature, the planet, and the cosmos must also be celebrated. Thus, the Pan-intrinsic person completes the circle. Or, as some Native Indians say . . . holds the sacred hoop.

Pan-intrinsic people are committed to a more global vision of holonomy. Central to their view is the unity between culture and nature. Their passion for harmony includes the Trans-intrinsic person's grace at aspiring toward and facilitating human harmony. The Pan-intrinsic extends that passion for harmony to the earth . . . the planet, and the cosmos.

A holonomic consciousness of nature provides the vision for the cultural involvement of the Pan-intrinsic person. Despite the role in which the person is involved, nature and culture are simultaneously present.

In the purest sense, each of the intrinsic postures has transcended ego. The Intrinsic has transcended personal ego, the Trans-intrinsic has transcended cultural ego, and the Pan-intrinsic has blended into a unity with personal, cultural, and natural systems.

I would hope that the reader, while trudging through these models, would bear in mind that *all* of the stages of personality spoken to by Harvey lived in all the ecologies of consciousness represented through the descriptions of Charles Reich. Also realize that there were discernible tendencies toward Consciousness III_R during Consciousness I_R. Neither the personality types 1_H, 2_H, 3_H, and 4_H, nor my labeling of these as Authoritarian (1_H and 2_H), Dependency/Manipulative (3_H), and the Intrinsic types (4_H), are distinct, unique, and permanently edged in fine lines.

Each person and each age represents a *tendency*. A *norm*. Figure VI shows a bell-shaped curve that represents the distribution of types at the time I started paying attention to these constructs, approximately fifteen years ago. It is laid upon the matrix of Harvey, Reich, and Samples. Note what it suggests:

Figure VI.

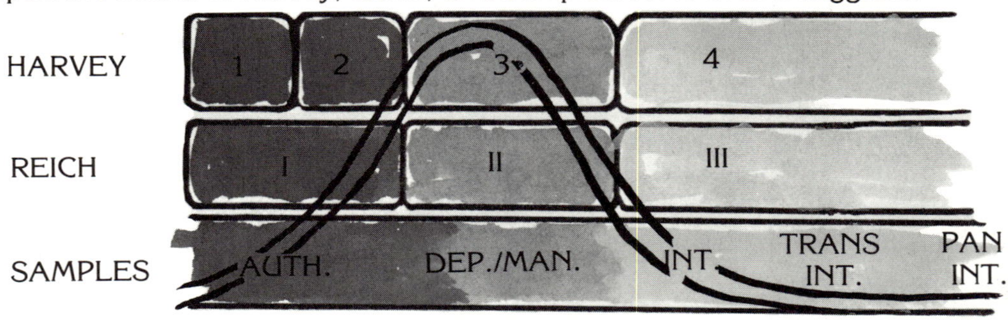

Figure VII.

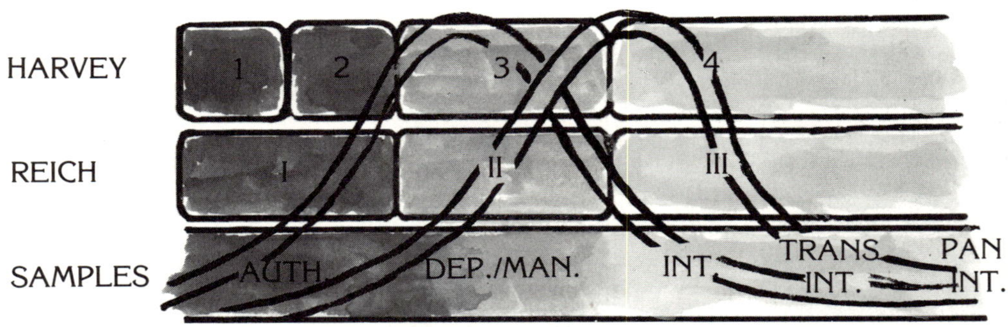

We could say that we were well into Consciousness II_R fifteen years ago. There was a high percentage of stages I_R and II_R personality types—in other words, a high percentage of Authoritarian and Dependency/Manipulative individuals. This curve would show a normative equilibrium between populations and overall social ecology. It is clear, however, that there were also small but noticeable numbers of people who were in the Intrinsic categories of Consciousness III_R and Harvey's type 4_H personality.

For me, Figure VII shows the configuration as it exists in the United States today.

At this time in the history of the United States, I feel that there has been a noticeable shift toward the boundary line between Consciousness II_R and Consciousness III_R. I believe there are significantly more System 4 persons than there were just fifteen years ago. Even if exaggerated in terms of specific numerical connotations, I believe the shifting tendency of the curve real and valid.

The basis for this belief is entrenched in the evolving patterns which—in sometimes trivial and sometimes profound ways—honor the intrinsic. The institutions in our culture trend toward more openness. Hospitals are commonly honoring alternative birthing techniques. Holistic medicine and its emphasis on wellness as opposed to illness is represented by nearly one-third of the program offerings at medical conferences. Schools trend toward more open and generalized modes of instruction. Churches are becoming more holistic and generalized in their commitment to their faithful. Again, no one has chronicled these patterns of change better than Marilyn Ferguson in her *Aquarian Conspiracy*.[4]

Perhaps even more significant than the widespread and quiet movement of the American people toward such directions is the nature of the simultaneous backlash. Just as I characterized some social critics as popularizing the "nostalgic backlash," so too do the most vulnerable traditions of society reflect similar backlash tendencies toward non-existent stability. The good old days aren't there any more. With the 1980 national elections now comfortably in history, the vacuous claims of "landslide," "mandate," and "return to stability" can be seen for what they are. In the face of *any* paradigm shift, there is a reflexive turn by bewildered people to the antithesis of what exists and is available. Copernicus, Newton, and Einstein all created such backlashes in the scientific communities of their time. Transformation toward the intrinsic holonomic postures is even more bewildering. It is leaderless, institutionless, and unorganized—and yet this shift is manifest in the staying home of nearly fifty percent of the registered voters during the 1980 elections. When no one plays the game, the rules change.

The return to rigid and rightist conservatism in the government, the blustering emergence of the Klan, and the amplification of the Moral Majority are all awesomely different reactions to that which has failed in American life. They represent pendulum politics, fascism, and zealotry, in that order. *Who* is in power is not the issue in contemporary politics. The issue is *what* is in power. A cultural trend toward the intrinsic cannot be led by Systems 1, 2, and 3 vision. Authoritarians, Anti-authoritarians, and Dependency/Manipulators are counter-evolutionary. They are counter to holonomy.

Perhaps their exaggeration at this time represents a message clearly documented in natural evolution. Exaggeration usually precedes extinction. Extinction is the result of closed-systems approaches in open-system settings.

It is useful to remember that to this point we have been drawing hypothetical curves that showed the shift from the mid-1800s to the present. These are supposedly normative descriptions of cultural tendencies in the United States. One could assume, however, that each curve is a normative personality/consciousness tendency in an individual. Figure VI (pg.106) shows a person whose main qualities represent Dependency/Manipulative with tendencies against convention. It is also clear that this person experiences *some* totally authoritarian moments as well as occasional flashes of intrinsicity.

If Figure VII (pg.106) represents this same person some time later, then we could suppose the person has begun to pay attention to those flashes of intrinsicity, and finds them liberating. This person moves him or herself toward a motivation/behavior style that favors such liberation.

Is this possible? As surely as evolution takes place for a phylum, it begins for the individual. I have watched hundreds change or evolve their basic or core personality. In most cases, it required that they pay attention to things within and outside of themselves that represented the more open-system contexts that existed. Once awareness was achieved, consciousness could follow. These are persons who accept Jantsch's notion of self-regulation.

The movement toward a more Pan-intrinsic or planetary consciousness is both a personal and a collective issue. As each individual ventures on this journey, so too does the human aggregate, or collective change its context. This is ever so subtle at first . . . but gathering in strength as it grows. It is well to remember that we are not talking of revolution here . . . we are talking of *evolution*.

The reflexive, historically infantile tendency of revolution to destroy the ecology that we disapprove of has never granted us peace. In evolution, ecologies are not destroyed. They are invited to higher levels of dynamic stability. And there are higher opportunities for self-regulation.

The movement toward the Pan-intrinsic or any of the Intrinsic postures can surely be affected, but the effect cannot be sustained. Those who deceitfully use planetary consciousness as their manipulative/dependency banner lose context rapidly. Many of these are Harvey's System 2_H types. With their exhausted moralities of revolution, they try to borrow from the mother—the planet—a focus for new-found reactionism. But these soon fail. The planet is too much an open system for their closed-system motivation. The assassin and the celebrator cannot bed together for too long.

Throughout the history of humankind, the planet has always been the sustainer. Out of fear and a withered consciousness, much of humankind has created an adversary relationship between nature and culture. This legacy is entrenched in the moral philosophies of the industrial powers. In both testaments of *The Bible*, there are nearly three-hundred references to the wilderness and nearly all are derogatory and pejorative. And yet, all visionaries of the past sought solace and solitude in the wilderness as they strove for spiritual insight.

Centuries of looking solely to other humans for spiritual nourishment and guidance is coming to an end. The evolutionary course of consciousness for both the individual *and* the culture is guiding us back to the mother . . . the planet and the cosmos.

PART III

VISION BORN WITHIN

Inherent within us is a genetic tendency toward survival. Chapter 9, *The Way of Life*, explores the implications and consequences of these inborn tendencies. This genetically endowed form of survival is not one of tooth and claw . . . not one of "closed system" thinking. Rather, it is one of open systems. It will require wisdom, openness, and a sense of harmonic blending.

Chapter 10, *Toward a Blending*, begins with a synthesis of the "way of life" and holonomy, and it continues to place this synthesis on the matrix of consciousness. It does so by exploring the notion that it is the *mind* which will inevitably recognize its "natural" harmony and blend those with cultural choices.

Chapter 11, *Whisperings Within*, creates a reference model for what many have called the transformation. I present the argument that *only* through honoring those innate human tendencies toward affiliation, empathy, and autonomy—can transformation and survival be insured. The distinction between true autonomy and the "narcissism" spoken to by closed-system critics, is made. The chapter ends with a descriptive model of the elements of the paradigm shift taking place in contemporary society. This final synthesis is drawn from history, biology, and psychology.

9

THE WAY OF LIFE

Life was probably born on earth over three billion years ago. In the countless generations of organisms that have lived and died since then, a remarkable rhythm has been put forth. This rhythm has to do with a kind of internal pacing that affects the blending between birth, death, and resources. When a creature is born, there is an almost instantaneous urge for nourishment. If the environment is rich, the newborn will flourish and grow rampantly until one day its food-getting is interrupted by an urge other than for food. To satisfy this new "consciousness," reproduction is necessary.

With reproduction, there begins a tendency to increase in numbers that put additional demands on food sources. As food diminishes, survival rate or reproductive effectiveness also diminishes. Whatever the causes, the final effect is a tendency toward the dynamic dance of ebb and flow spoken of in Chapter 8. Although there may be short-term deviations, the population of creatures, their food supply, and natural enemies come to a kind of harmonic stability expressed best as a dynamic equilibrium. Dynamic since it evolves in an open system, and in equilibrium because it has continuity.

Most of our metaphors of equilibrium are static or closed-system examples. Teeter-totters, the scales of justice, and the stack of blocks in a child's play area serve. In fact, most members of our culture perceive equilibrium as being a state of fixedness. I believe this arises primarily from the ever-present examples of equilibrium drawn from technology. Strangely, technological equilibrium results from interaction. The outcome is that equilibrium results from humans taking action on some system and bringing it to some kind of closed-system balance. The very laws of thermodynamics have nurtured a false security in the minds of most observers.

As a result of this tendency, the open systems of nature have been looked upon by closed-system minds—minds closed by the metaphors of machines (thermodynamics) they use to view the open world of nature. The organism, its food supply, and reproductive urges are often viewed as a closed system. Technocratic humans treat nature as if it were an internal combustion machine, and they owned all the fuel.

To the technocratic human, any increase in numbers should be balanced by an increase in food supply. If this does not occur, then intervention follows and the organism, if it is favored like cattle or sheep, is provided new or supplementary food supplies.

If the organism in question is not favored, then intervention takes the form of an intentional diminishing of the food supply (use of herbicides), or direct killing of the creature (pesticides, etc.). When such action is taken, the interventionist is usually surprised to discover how much intervention is necessary to accomplish the task. The technocratic human becomes amazed at the sudden burst in reproduction that takes place when the food supply is threatened or when the population is artificially limited. It doesn't make sense, mutters the interventionist, who then redoubles his or her effort.

At this point it may seem that I am ready to single-out agriculturalists, cattle and sheep raisers and other contemporary villains. Many of these rightly deserve chastisement but the "protectors" of nature are also as guilty. Wildlife biologists, fish and game managers and a bevy of conservationists are also prone to quick-draw interventionism. Witness the conservationists' reaction to the crown of thorns starfish on the coral reefs surrounding Australia. Millions of these starfish appeared and began to devour the living coral. At first the cry was "worldwide pollution" and great worry emerged about what to do. Uncounted employment hours were spent puzzling over which kind of intervention would be most appropriate. Years after the first concern, the quibbling still went on. This was fortunate . . . for now it appears this bloom of coral-killing starfish is simply part of a cycle no one had had the chance to witness before. *Any intervention would have been a corruption of the cycle.*

The pity of the interventionists' "wisdom" is that as intervention is increased in the name of equilibrium, the system becomes more and more closed. In other words, the attempts to establish equilibrium in open systems immediately introduce closure. Continued intervention is required to maintain closure. Regardless of how much effort is expended to maintain a closed equilibrium, it will inevitably revert to an open system.

The way of life is to continue. Patterned tendencies toward continuity are innate. Any attempt by intervention to alter that continuity reawakens shards of the innate patterns and new options emerge. In the presence of such induced stresses, the subtleties of evolution are called again into play. Evolution is nature's way of continuity.

In his book, *The Survival of the Wisest*, Jonas Salk says:

> As nature continues its game of biological mutation and selection, and as man plays his own games of selection of ideas and of cultural innovation, nature will have the last word.[1]

The inherent tendency in life is to continue in an open system of dynamic harmonic equilibrium. The outcome of this tendency is simply called *survival*! As one looks upon the graphic expression of the tendencies toward survival of any group of organisms in natural systems, there are seen to be cyclic fluctuations in their periods of abundance and scarcity. Such a curve is expressed in Figure VIII.

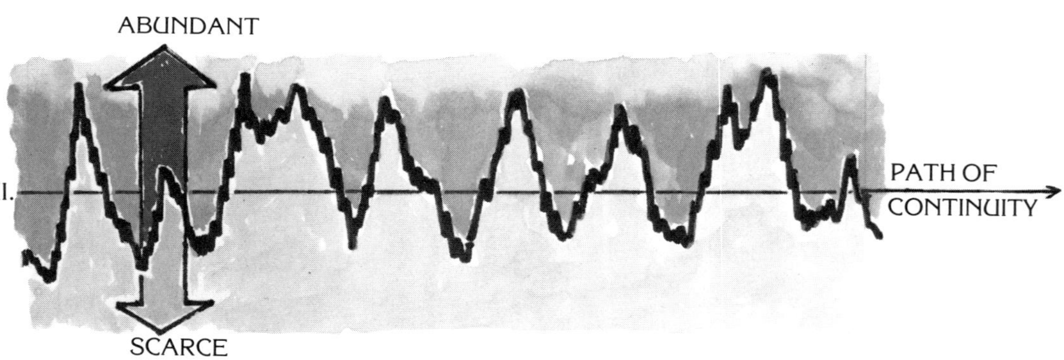

Figure VIII.

These ups and downs represent the focus of study of hundreds, if not thousands, of natural scientists the world over. They represent a kind of natural rhythm of life, death, and continuity. Within these group rhythms are much more minute pulses which pace the inner songs of survival. These are genetic. They are the patterned pulses of holonomic continuity.

Am I forgetting extinction? What happened, long before human intervention, to the creatures that became extinct? Extinction appears in our science books like an event . . . like an instantaneous scene shift in a wide screen movie. Not so! In nature extinction must be seen more as a testimony toward transformation than toward termination. All species of a particular genus never become extinct. The genus continues. Holonomy may be manifested more in the whole of life than in any specific branch of life.

Those who look upon nature as might a technological voyeur often forget the brevity of human life span. We can seldom watch the undulations of life and death of other organisms for more than a few cycles. Had we the potential for extending our vision, it would become clear that there is a larger, more persistent pattern being followed. This larger pattern is expressed in a *sigmoid* or "s" shaped curve. The development of such a curve's choreography of survival is shown in Figure IX:

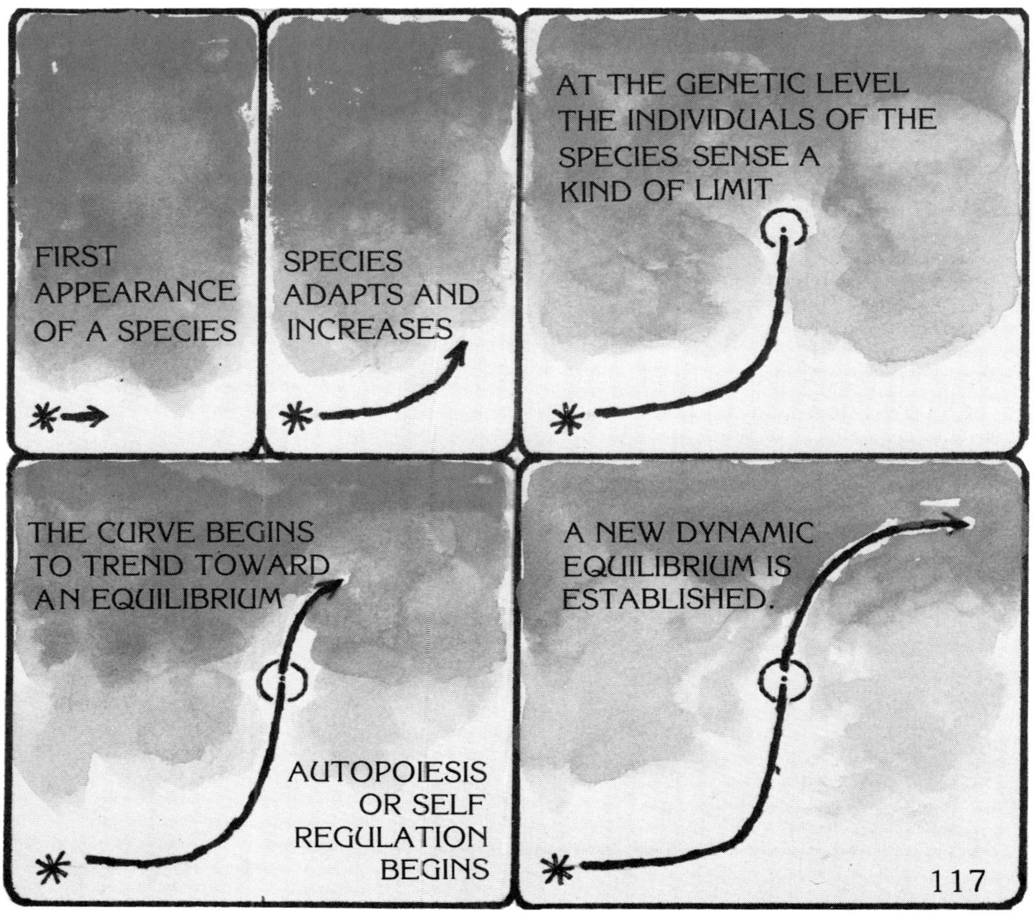

Figure IX.

Figure X

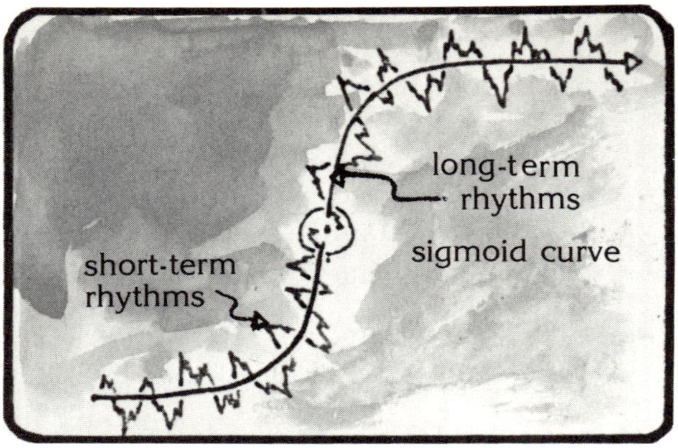

Jonas Salk, quoted earlier, describes these processes in detail in his brilliant book, *The Survival of the Wisest*. Salk believes the sigmoid curve to be the standard expression of the development of equilibrium in the processes of evolution. He even suggests that the pattern necessary for harmony is genetically encoded in all life. The caprice of time has caused many of us to focus upon the rhythm of Figure VIII rather than the sweeping grace of form and concept in Figure IX. Perhaps a combination of the two would make this clearer.

The way of life can thus be expressed in two scales of function. There are rhythms of short-term periodicity and, there are also longer, sweeping patterns, which move toward an equilibrium in open systems.

There are many who perceive themselves as the alarmist, potential saviors of life on the planet. These doomsayers are disturbed by Salk's claim that this process is part of the basic, innate patterning of life. It is discomforting for many to be confronted with the possibility of this harmony being natural. This disquiet group sees the capability of nature running itself as being a negation of their interventionist ego. This school of "catastrophists" is quick to point out another curve. This doomsday curve has been made popular in the decades since the Malthus essay on the principles of population in 1798. Its curve is show in Figure XI:

Figure XI

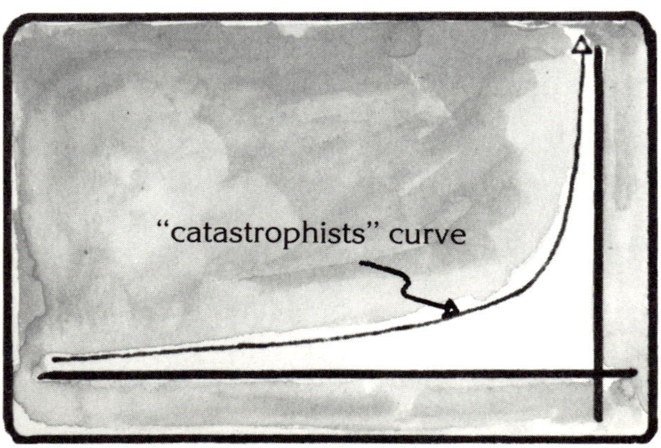

This curve, say the catastrophists, is a truer depiction of the way things are. This view is, in effect, a return to the economic depiction of supply and demand that was represented metaphorically by the creature mentioned at the opening of this chapter. To live, to eat, to reproduce, is surely basic to life. Going on, the catastrophists seem to believe that when left to themselves these patterns get out of balance. Fearfully, they see the exponential curve shown in Figure XI as an expression of what happens when humans don't keep things in check.

But what the catastrophists seem not to trust is nature. They appear unable to realize that if there is the ghost of planetary consciousness, then it is likely that humans have the capacity to begin trends toward larger-scale equilibrium. The human organism is a newcomer to the planet. Only a half-million years ago, *all* humans were hunters and gatherers. Little more than forty thousand years ago humans began to travel and colonize. About twelve thousand years ago, the dog and goat were domesticated and the first species (a small horse) was made extinct by humans.

Up until then, natural consciousness and natural equilibrium seemed to be achieved, but surely there were cycles of scarcity and abundance, with good and bad times experienced by all.

But then, ten thousand years ago, the first seed was planted. Agriculture began. This was the first major intervention of cultural consciousness over natural consciousness. Now food could be grown and population growth could be managed. The church, the school, and the military would all be invented in relatively short order. They would add reason, substance, and power to the minds of those the grain favored with survival.

Two thousand years later came the wheel. With it, natural planetary consciousness would become systematically suppressed by nearly every culture on the northern half of the planet.

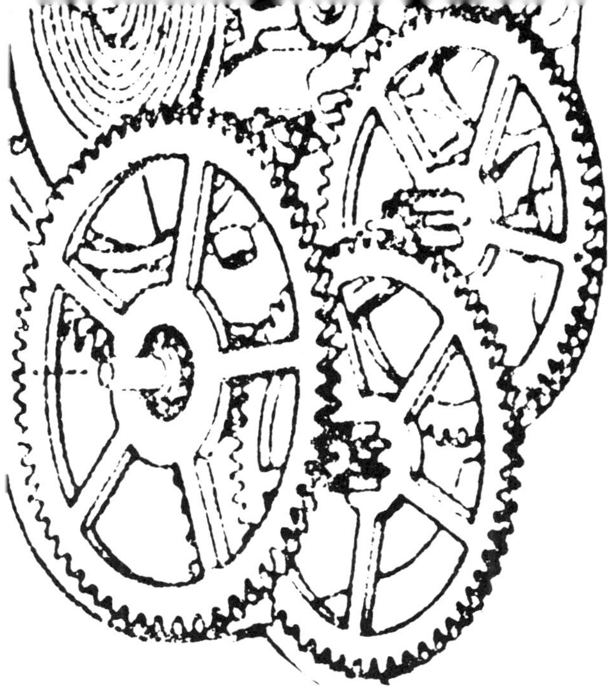

It was intervention that nourished the runaway contexts of the curve. Medicine kept people from death. Trade and commerce shifted goods about to supply the demands of a technologizing world. The intervention became formal in law and politics and created a cultural context so distant from nature that one had neither the need nor the consciousness to re-establish a bond. Nature was a resource of things to be used. Its physicalness was its virtue, and that was all there was to it.

But is it?

What seems to be awakening within the patterned energy in our own being? What is it that seems again to be resonating with the planet? Have we been out of synchrony for so long that evolution is inviting us back, back into the patterns of harmony? The coalescence of ideas from such people as Margaret Mead, Buckminster Fuller, Rolf Edberg, Vine Deloria, George Leonard, Jean Houston, Joseph Campbell, Jonas Salk, Theodore Roszak, O.J. Harvey, Marilyn Ferguson, Charles Reich and countless others is sculpturing a new consciousness. This is not a consciousness devoid of cultural essence, but one that seems capable of entraining to the music of the planet, as well as the music of the city. The catastrophists and their acerbic vision, strangely enough, are perhaps one phalanx of this innate human tendency. Unfortunately, their dourness and near fascistic tendencies often represent a less than hopeful vision of humanness. In Harvey's terms, they are too often seen to be System 2_H. They emerge as Anti-authoritarians, whose effectiveness is vulnerable to being swallowed up by the very system they wish to overthrow. For them, the new consciousness is read only as innate urges toward more closed-system intervention.

It is the way of life to continue. As humans, we are part of that continuity. We, as a species, have intervened mightily and are destined to continue our intervention. But if we are at the inflection point, if we are at that threshold of consciousness that comes persistently from the very pattern of our being, we must choose. We must choose between a dynamic movement toward harmony and blending and a movement of discordant, reflexive conflict with the static, closed-system dominance of equilibrium-focused culture. Only closed systems reach equilibrium. In the words of Jantsch, "Equilibrium is the way of stagnation and death."[2] But he means the closed system equilibrium of the closed system thinking interventionist mind.

The most closed among us are closed to themselves, closed to culture, and closed to nature. When such people feel the presence of a holonomic invitation, they will respond most often from the expertise of their psyche. They will respond in closedness. It is a frailty of humans to see only what we know. It took us so long to discover the black holes in space because we always looked where the light was better . . . at the stars.

To regain the openness that will honor what Salk writes of, we in ourselves must relinquish the closedness and fixity of our cultural vision. We must look anew at the planet. We must try to see nature again, not just as a physical resource to satiate our machines; rather, we must learn anew how to see the spirituality of nature.

To do this, the Authoritarian and Dependency/Manipulative consciousness that have dominated the past must give way to the Intrinsic. Consciousness I_R and II_R of Reich must drift into the background of options so that a more fulfilling expression of human culture can emerge. These ideas invite us back into knowing the way of life, of evolution, of harmony. As we regain these forms of wisdom, the culture will transform. It will transform at the most basic unit of itself . . . the person.

As each of us addresses the inflection point of our *own* equilibrium, we have a choice. We may choose the equilibrium of death and dying, or of harmonic evolution. If we choose harmonic evolution, we will enrich the culture that we hope to achieve. The enrichment will come from the reopening of closed systems within ourselves. Perhaps humans are the only organisms on the planet that can willfully slow down and inhibit their own evolutionary process. We can create, by choice, closed systems to effect this inhibition. But as surely as we can choose to shut ourselves away from holonomic consciousness, so too can we choose to re-enter.

Walt Whitman said it best when he wrote:

> There was a child went forth every day
> And the first object he looked upon
> And received with wonder or pity
> Or love or dread
> That object he became.

So too we must become that which has hope . . . that which has the openness of nature. Nothing will be gained if we become more closed, more parasitically allegiant to the very systems that would destroy us. The restoration of planetary consciousness requires that we *each* be fulfilled. The human we *are* is the human we see. Hollow humans cannot invite us back to the mother.

10

TOWARD A BLENDING

As Pribram, Bohm, and others, including myself, pursue the concept of mind as a holonomic consequence, we are each, in our own ways, awestruck by the simplicity of the context. The brain can be viewed as an organic pattern of energy in the frequency domain; it is pulsing, vibrating, entraining, and becoming dissonant and re-ordering again. The mind, an extension of the holonomic harmony, develops from both within and beyond the brain. It can be viewed as a pattern that blends in and out of context. It breathes itself into being, as George Leonard's "silent pulse," and as Bohm's "clouds of enfolded presence." Mind, brain, matter, and energy are all part of the same folding in and enfolding rhythm. The brain, that substantive part with organic unity, creates its holonomic patterns through flickering configurations of thought. The brain, the body, the person—all of them combine in a temporary configuration which, though shrouded by the mysteriousness of the holonomic pattern called life, persists. Once the *life* pattern enfolds back into the cosmos, then *life* in this configuration ceases . . . but other energy patterns go on. The stuff of which we are made moves out of the *life* configuration and blends again with the cosmos.

If we can entertain the idea that matter is a holonomic pattern of energy, then we can describe *life* as a special case of unique harmony among the families of configurations called matter. This *life* harmonic organizes and creates a substantive kind of persistence to the organism. The organism, in turn, tacitly and explicitly begins to obey the evolutionary patterns of change we call growth and maturation.

Viewed this way, if one were to fall from a cliff or smash up an automobile, one might create a shocking disharmony which might, in turn, destroy the *life* harmonic. Disease is a far slower and dissonant presence, and it may take months or years to disrupt the life harmonic. But it, too, may destroy the harmony. From this perspective, we may state very simply that illness is holonomic dissonance while wellness is holonomic harmony.

Where do such flights of fancy take us? Our heritage of Newtonian vision has trained us to seek discreteness in events and to search out the causes of effects. Such a heritage has also taught us to be myopic to the four-dimensionality of patterns. We have been trained to perceive events, which, in the Newtonian world, inevitably have cause. The linkage of an effect to a cause is a primary statement of closure to a reductive thinker. Strangely, this bias affected the way we thought of mind, brain, body, and person.

Many in the past have gained solace from the reductive models of brain and mind that have preceded us in this century. Let us review briefly. In the early electrical probing of the cortex, it became comforting to know that stimulation of one part of the brain would result in a specific mind-body linkage. Taste, smell, memory, and vision—to name a few—had specific loci in the brain. Touch this area with an electrical stimulation, and one got a specific mind-body response. The brain had a neural kind of geography.

The students of the lateralized functions of the brain followed with a promise of new certainty. Their initial claim was that the left brain did so and so, and the right brain did the rest. Metaphorically, it can be said that there was a division of labor in the brain. After a decade of research, it is clear that *content* is not lateralized, but ways of processing content may be. Also, the more closely one looks, the more it appears that the whole brain is excited during each learned experience. Moreover, it also seems that the whole brain tends to respond when experience is resurrected in memory. The closed-system brain was losing ground.

Now, with the open-system context of the holonomic brain, a newer view of knowledge and brain-mind function is afforded. We can begin to think of the mind as primarily a brain-generated pattern that configures a holistic experience, rather than as a single element thought (Chapter 3). When this "pattern" is generated, it is often economical to pay attention to only one part so that one can remember "the" answer. For example:

> I had recently misplaced my fishing license, and it had been over a month since I had last used it. That occasion had been on our five-year old son's first camping trip. The friend we had gone with was an expert camper. Our son and I fished before we went to set up camp. I had put the license in my shirt pocket and was glad to have done so because a game warden arrived and checked its validity. My son watched with interest. He devoured the whole scene. He wanted to know what a fishing license was, who the man was, and why our worms were dying. All this against the background of chatting with neighboring fishers. We caught one fish which he shared joyfully with all passersby. Later, we camped and fished some more, and had a great two days of it. But now—a month later—I couldn't find my license. We were about to go fishing again, and I was mildly agitated. I looked everywhere it *should* be, but to no avail. Against this background, I was confronted by our son, who asked, "What are you looking for?" "My fishing license," I answered, while I continued to look. He fell silent and lapsed into a daydream expression. After a few moments, he shouted to me, as I had gone into another room. "It's in the wallet with my picture on the front." Even in my agitation, I stopped and asked what he meant. I had no wallet with his picture on the front. I looked through my regular wallet again. Still no license. He said, "Not that one Dad . . . the one with my picture on the front." Then it struck me . . . my professional date book and calendar. On its cover, I had inserted his picture. It was hardly a wallet, but I looked in it. There was the fishing license. It was totally accidental that I would take my calendar on a fishing trip, and absolutely unseemly that I would insert my fishing license into it. Even after retrieving it, I couldn't at first remember having put it there. I asked the child how he had remembered. He said, "I imagined it all back and I saw you doing it." He told me about "pieces of the picture" and how he was seeing the dead worms and knew we had to get some more and that I would need my wallet to pay for them. He then remembered that I had slipped the license from my shirt pocket into the date book calendar which was lying next to my regular wallet in the camera case.

The amazing part of all this was my embarrassment as I realized how the whole scene that our son Stician had explored was also available to me. I was eventually able to re-create the entire scene for myself as well as an array of others. I explored in my own consciousness the selective way that I had scanned these resurrected contexts in my mind. The wallet, the pocket, and even the glove box of the car kept dominating my search. They did so because they were *logical. They made sense*! My probing consciousness was dominated by an internal set of cause-effect perceptions.

Stician didn't have these biases. Therefore, in *his* exploration of consciousness, there was no predisposition to make sense. Suppose then that a thought is a holonomically-patterned form of energy . . . a pattern with lateral and vertical extent. Few of us have been trained to see the *whole* pattern. We have instead been trained to look at the pieces. If I have been trained to see pieces, I will believe reality is a collection of them. I may then, as I have done, explore my mind for four decades and come to the realization that it is an aggregate of pieces.

Memory, knowledge, wisdom, science, poetry, and art are all pieces of reality. With such a mindset, I can go through life and never have the ghost of a notion that there is an overriding pattern. The child, on the other hand, is not yet predisposed to pieces. Stician can see the whole and has no logic to uphold, other than the logic of experience. *He can look anywhere in the pattern, while I have limited my exploration to those elements of the pattern I have experienced before.*

There is a new meaning then to Karl Pribram's statement that "consciousness is what we pay attention to." It follows that to extend consciousness (chapter 4), we simply need to pay attention to different things. It is my history that determines what I know and what I perceive. And although it is true I am sculpted by my past, I may always decide whether or not I choose to continue in that imprisonment. Little Stician fished freely in his entire mind, while his father fished in a file folder of historical logic.

I believe that Pribram's phrase "paying attention to" means simply *to develop harmony with*. This may be one of the great gifts of C.G. Jung, with his sponsorship of the concept of synchronicity. Harmony, synchronicity, entrainment, vibrations, resonance, and blending are all common metaphors in what Theodore Roszak calls the *human potential mosaic*. When there was a human potential *movement* in the 1960s and early 1970s, the metaphors of this movement were intended to be more holonomic. Yet learning to "relate," "go with the flow," and not allowing things to be "laid on you" had their own hazards.

The political-personology of the encounter-group focus of the 1960s always runs the risk of making manipulation and tacitly-expected affiliation the final state of consciousness. They may well be best exemplified by *est*. While many have used their *est* experience to expand their consciousness, some have chosen to use it as a means for more effective manipulation of others. Still others have turned their experience into a mechanism of servitude to *est*, one which is exemplified by volunteer work for *est* or a powerful tendency to proselytize. *Transformation can never be guaranteed from the outside.*

The 1960s represent less of a transformation than a reformation. Models of human being emerged that concentrated on the interpersonal. Consciousness became inter-human consciousness and, only late in the decade, did recognition of transpersonal qualities emerge. (I am intentionally leaving out the drug-induced consciousness contexts of the 1960s and 1970s, as I have yet to be convinced that they contributed to natural consciousness.)

As the transpersonal search began, much of its focus was oriented toward the attempt to introduce esoteric practices and disciplines to the main vectors of humanistic psychology. Yoga, meditation, and T'ai Chi were among the first, and with them came a more widespread acceptance that you can *do* things with your body that awaken new vistas in mind. This new use of old disciplines was frequently and unfortunately accompanied by tendencies toward affectation. But again, recall Roszak's words on page 92; he points out that the tendency to explore consciousness through the practices and wisdom of other cultures is far more important than the few offensive, turbaned, and white-robed miscreants who prostitute the enterprise.

For me, the context of the transpersonal is far from its adolescence; it is still in its infancy, and its promise is before it. One great advantage of the emphasis on the meditative and yogic traditions is that these clearly predate the consciousness tradition embedded in the Western, technocratic Judeo-Christian view. The advantage of this antiquity is that its roots were closer to pre-literate humans . . . closer to people whose basic affiliations were with the planet.

Consciousness is intentionally being extended by dozens, if not hundreds, of efforts in our own culture. To those who look about in the fabric of contemporary American life, it is clear that the reformative human has evolved toward the transformative human. There is an expressed yearning for a unity, not only with other people, but with larger systems as well. Consciousness of this extension is appearing in sources as diverse as franchised weekends and American Medical Association meetings, in the lyrics of John Denver's music and the pulpits of urban cathedrals, in advertisements by Weyerhaeuser and the penetrating planetary stories once presented by Walter Cronkite and uttered before his, "That's the way it is."

It is this persistent and growing presence in our society that led Jonas Salk to express a second meaning to the inflection point of the sigmoid curve (Chapter 9). Recall that the first part of a sigmoid curve looks just like the doomsday curve of the catastrophist. See Figure XII.

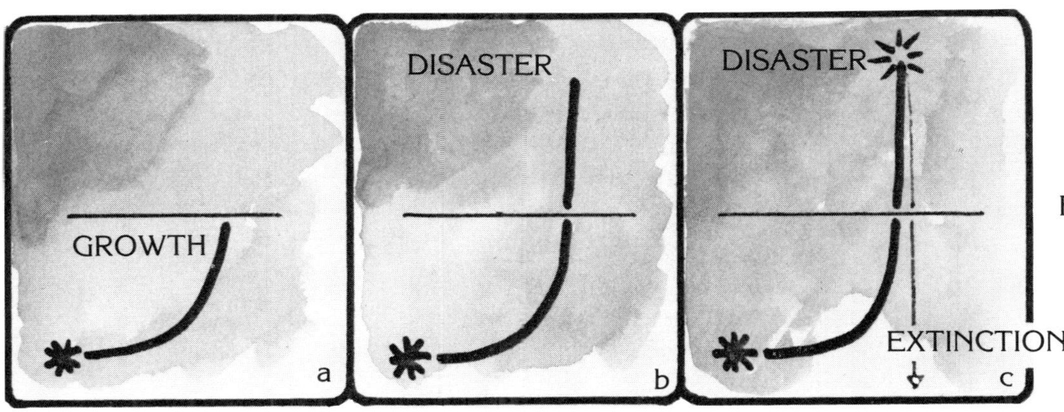

Figure XII.

The catastrophists see the relationships as seen in the three stages of Figure XII. (a) A growth as resources are utilized. (b) A runaway period where greed, exploitation, and profit excesses create the conditions for disaster. (c) And finally, the inevitable disaster that results in annihilation or extinction.

Recall that Salk emphasized that the curves of organic origin, growth, and stabilization are represented by the s-shaped or sigmoid curve. It is characteristic of this curve that in the early history of a population of organisms it seems to be sweeping upward with the runaway exponential qualities of the doomsday curve. But then the population reaches a place where the exponential nature of the curve is abandoned. This point is called the inflection point. It is the point at which the curve changes its characteristic from exponential growth to diminished growth and eventual equilibrium.

Salk speculates that the inflection point is genetically or, in my words, holonomically endowed. We can begin to reach an inflection point in the pattern of our consciousness as a population, a point at which *the individual and the pattern develop a mutual resonance*. At that point, both the individual and the population pattern begin a gradual entrainment that results in survival.

Does Salk mean that our genes turn off our reproduction, our exploitativeness, our tendencies toward growth? I think not. It is not as though the genes have specific pre-coded actions, but rather that they are encoded with the possibility of an evolving consciousness. No one can expect a sudden physical burst in evolutionary structure or function. Instead, it is likely that by using the qualities of consciousness that we now possess, humans will seek to create personal consciousness and its attendant cultural expression that nurtures stability, continuity, and quality life.

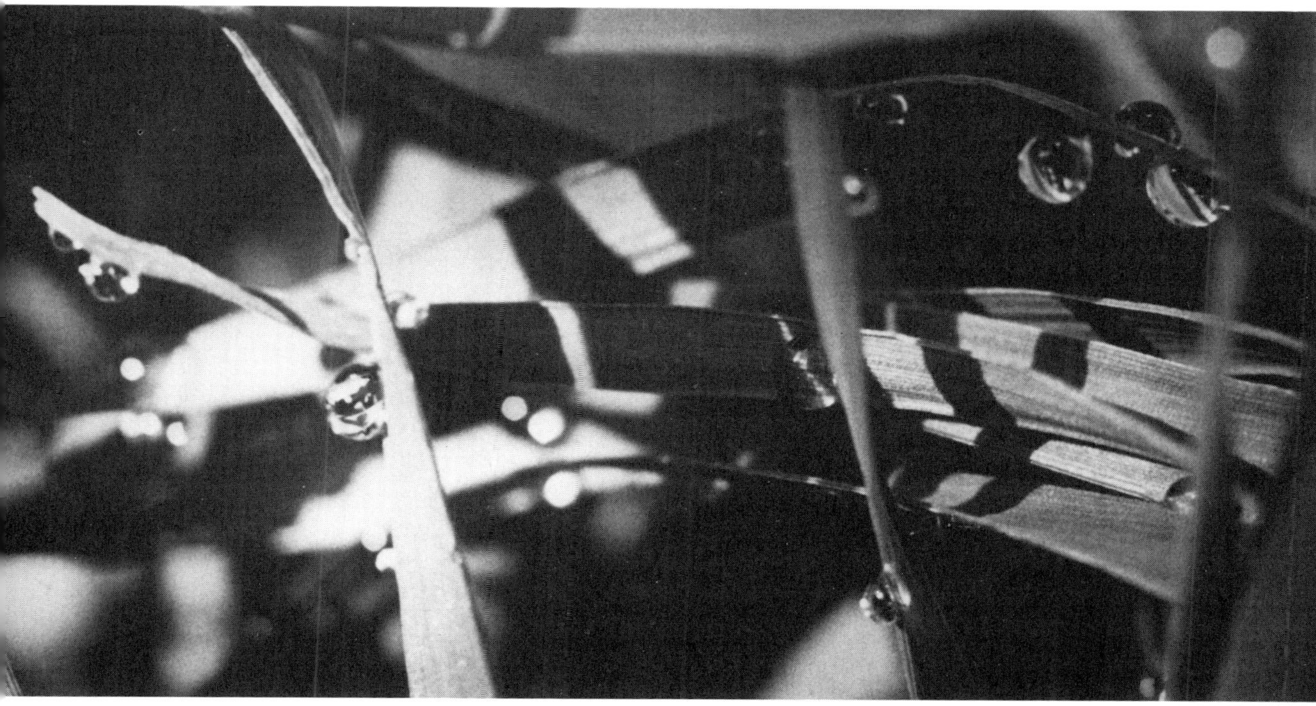

Perhaps the sociobiologists' arguments supporting staunch genetic control might be viewed as an expression of contextual consciousness. It is possible that the reflexiveness to act *is* in the genetic make-up, but the options governing actions are more closely linked to consciousness. This means that while humans may well be "programmed" to respond, the nature of the responses may be part of an evolving array of options which are configured by evolving qualities of consciousness. Sexism, ageism, racism, and all the other genetically-reflexive characteristics described by sociobiologists may in fact be historical descriptors linked to the early stages of evolving consciousness, and not at all mandated by our genes.

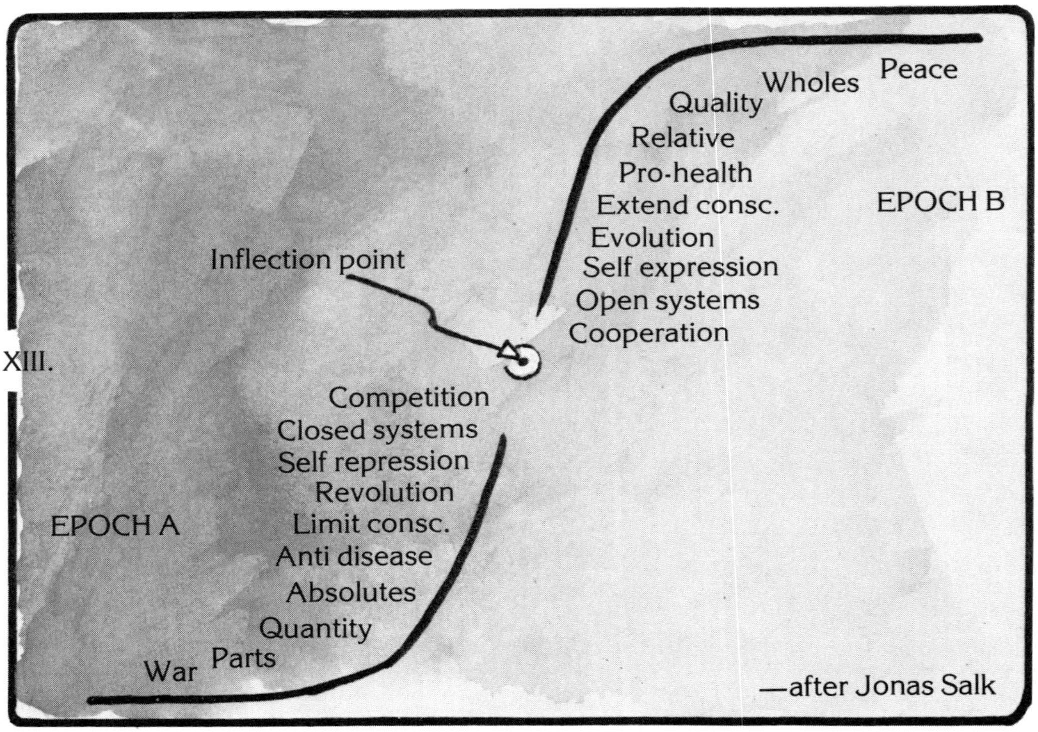

Figure XIII.

Figure XIII shows a hypothetical sigmoid curve expressing the possible shift of species *homo sapiens* from exponential growth toward stability and continuity. Salk labeled these Epoch A and Epoch B. Epoch A is characterized by the cultural values that we have represented in the past. He sees the inflection point as being at hand. The transformation of values from Epoch A to B will see a shift represented by the word pairs in Figure. XIII . . . movement from competition (a cultural trait) to cooperation (a natural trait). Quantity consciousness will shift to quality consciousness, and so on.

Salk goes on to suggest that to achieve this, humans have to learn to accept *all* parts of their mind, the intellectual and the intuitive. The lateralization adherents would say both the right and the left hemispheres. In holonomy, we might say the whole of the pattern.

Salk's wisdom, expressed nearly a decade ago, is being substantially nurtured by contemporary futurists. Peter Schwartz, in *Quest* magazine, cites the role of individual consciousness and its effects on counterproductive institutions.[1] Schwartz believes that the reversal of and disappearance of colonialism in the thirty-five years since World War II was the result of the change of the *idea* of colonialism. Colonialism vanished not because of revolution or force of arms; it disappeared because it was an expression of repressive consciousness. Likewise, Schwartz believes that the near disaster at Three Mile Island has affected the consciousness of millions of Americans in regard to the legitimacy of nuclear power.

Willis Harmon, the noted futurist director of the Stanford Research Institute, is particularly articulate in regard to expected changes. In personal communication, he cites a remarkable list of evidence for the transformation in consciousness.[2] Among these are:

- increased awareness of contradictory dilemmas
- reaction against exploitation of environment
- reaction against the secularization of human energy into the economic system
- reversing the secularization of knowledge to technocracy
- role of mass media

These are but a few of the indicators that are growing daily in the matrix of modern American culture. The media, through documentary and fiction, are addressing issues formerly untouchable. The machinations of politicians are now televised and taped in a network created by themselves, and we have all become privy to the dignity and inanity of the technocratic law. Francois Revel saw it in his study of American life, *Without Marx or Jesus*, a book that was among the first to document a shift in consciousness in America.[3] Revel saw the breath of openness emerging from the closed box of American traditions. Those who built the box cried "Pandora," and tried to discredit those whose songs of consciousness swept across the ages. Walt Whitman sang holes in the box, and was called a deviate and pervert. Thoreau sang loudly of the morality of civil disobedience, and was branded a criminal. Revel heard the effects of these songs and called them the greatest revolution humankind has witnessed . . . *and it was leaderless*!

Remember that consciousness is what we pay attention to. If Revel's "revolution" exists, then we can see it by paying attention to it. Our own minds move inexorably toward contexts of openness. We are being drawn by resource and spirit back to the patterns of nature. The limits of our exploitation may well be emerging, not because we are out of petroleum, but because we are choosing to be less exploitative.

When Salk's Epoch B gains in definition, it will do so because we *each honor the values inherent within our genetic mandate as homo sapiens*. We as individuals will serve to define it. Salk invites us to be alert to the cues of transformation from *ego* survival to *being* survival. *Ego* survival is survival in the present. *Being* survival seems to move toward a "destiny not altogether discernible."

I would propose that this destiny is the persistent and spiritual re-awakening of planetary consciousness. So strangely complex, and yet so simple, is the realization that from a base of exploitation we learn to exploit. From a base of celebration, we learn to celebrate.

11

WHISPERINGS WITHIN

I believe that what has so far been proposed is possible, that humans are genetically endowed to survive, and genetically endowed to transform. Although social ecologies affect us powerfully, people can intentionally transcend closed-system ecologies and create within themselves an open-system personality. Upon all of this, let's suppose that evolution influences both the person and the ecology in a dance of interrelatedness.

If, as Revel and Ferguson say, there are no leaders in this evolution—then where must we look to gain guidance in this proposed emergence? I suggest that the place to look is within, inside the being that we are at this very moment. I propose an exploration of self. I propose an entering of Jantsch's context of autopoiesis . . . self renewal.

A prevailing backlash to the past two decades of personal introspection—one that has been labeled the "new narcissism"—had its seeds planted long ago. In the very beginnings of the gradual shift from the rudimentary harvesting of berries, grubs, and seeds to pursuit hunting a higher state of organization was required. The running down of fleeter game required the kind of teamwork that enforced affiliation half a million years ago. Agriculture and surplus food production represented a quantum leap in creating contexts of highly organized affiliation. The end of this progression can be seen in contemporary tendencies toward all aspects of the military, religion, industry, commerce, and the entire fabric of technological society.

In the late 1800s, Emile Durkheim, in a study of suicide, came to the conclusion that more people killed themselves who were not affiliated with others in an organized way . . . ones who had not established a personal kind of servitude to group life. An outgrowth of Durkheim's studies was the growing field of sociology and one of its divisions, social psychology.

A part of Durkheim's model relates to a three-part composite of what constitutes the fulfillment of human need. The three qualities that were necessary were *affiliation*, *empathy*, and *autonomy*. Moreover, Durkheim saw these in this order. First and foremost to a newborn and immature human was the necessity to join the human community by affiliation. Bonding, family ritual, and socialization all represented various "normal" aspects of this process. As children acquired experience and learned adherence to the rules of affiliation, they were supposed to develop the capability for understanding the common *good* of approved emotions, relationships, and interdependence. As this *good* became recognized, the affiliated person could begin to empathize. In a sense, empathy was a shift from a base of obedience and acceptance to one of understanding and projection. Empathy was a state of being which insured that the rules and rituals woven obediently into the fabric of one's own psyche would be recognized appropriately in others.

The last of Durkheim's stages was autonomy. Autonomy is defined as self-reliance. If a person became acculturated, and conformed to the rules of affiliation, and later developed the skill of seeing the same suite of conformities in others, and, furthermore did not deviate significantly, then this person could be allowed autonomy. Durkheim effectively reserved autonomy for the stages of life when the individual was so imprisoned by the accepted customs of a culture that it was impossible to immerse in selfhood without entering into a psychic hall of mirrors reflecting only one's past. In a true sense, autonomy defined as self-reliance was a charade. In fact, freedom postured for this pseudo-autonomous person was premised on the culture's assurance that the individual would not step beyond his or her own conditioning.

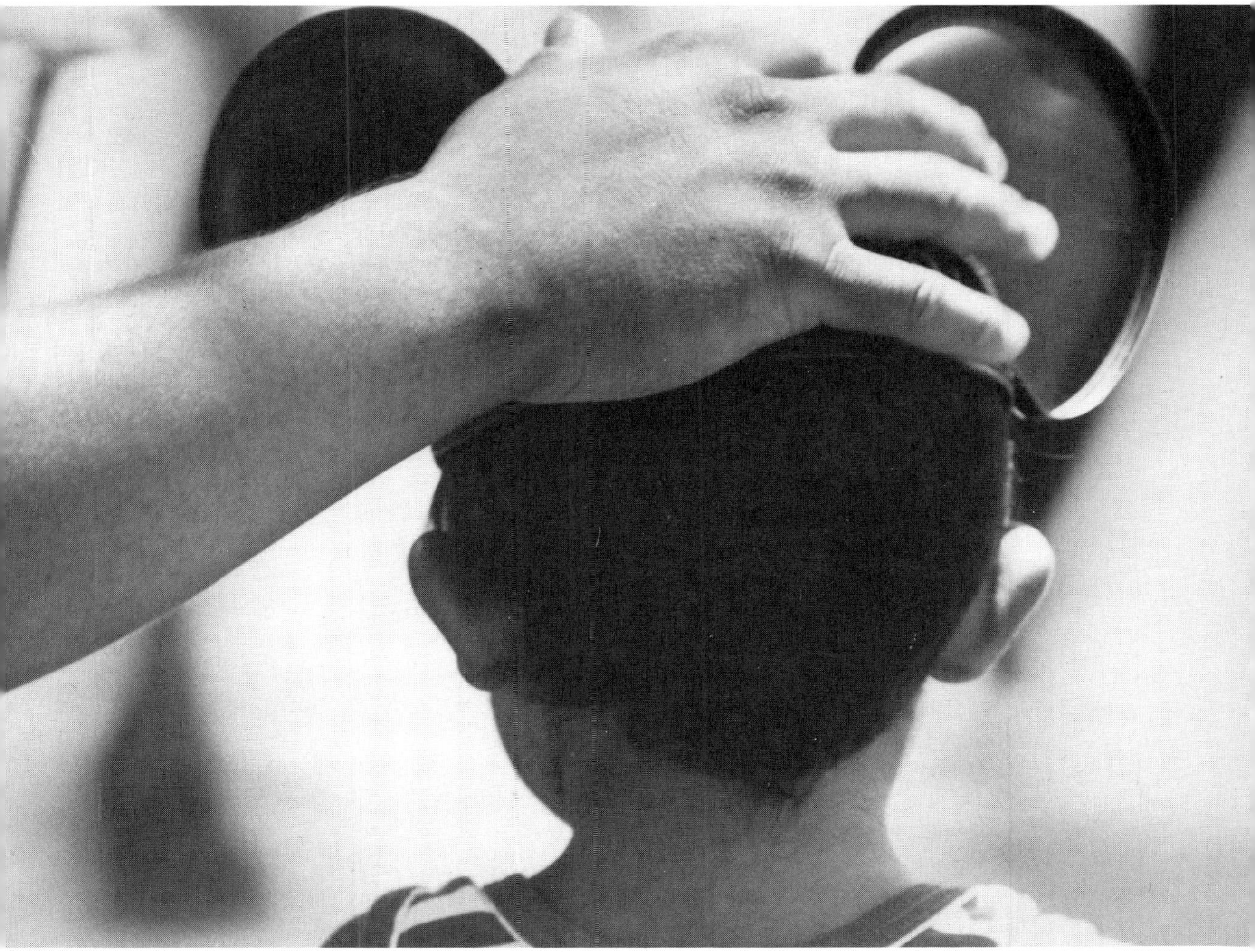

The "safety" of such an individual would later be explained by B.F. Skinner as operant conditioning, a technology in the philosophy of behaviorism. But this was the womb for the personality that O.J. Harvey would call System 1 and 2. The society would be labeled by Charles Reich as Consciousness I.

The "autonomous" person in this distorted perspective was truly a pawn of culture. It is such distorted perspective that creates the fearborn label of narcissism. Humans in contemporary culture recognize the flaw in this classic definition. Or better said, the definition, self-reliance, is fine; it is the premise behind it that is flawed. People the world over are beginning to recognize that there is perhaps an intrinsic worth to the human that transcends the cultural descriptors. Everywhere, people are recognizing that when the tyrannical side of cultural confinement is in direct conflict with the intrinsic person, then, most often, all the person needs to do is walk out of the prison. Be assured this is not easy . . . yet it is far easier than most would believe.

Let me pursue this idea. When first a "professional"—as designated by my cultural training and achievement—I became aware of a suite of expectations that culture had about me. For example, there was an expectation that I would dress in a manner "fitting my station." A tie was expected. I wore a tie.

Through being a professional I gradually met others. Some of them did not wear ties. At first I greeted their presence with a flood of insecure embarrassment. It was as though their flies were unzipped or bra straps dangled out of a sleeve opening. Our initial time together was wasted because I was so anxious about who would—and when they would—be discovered. I expected them to be banished.

After a while, when I knew they were not going to be banished, I began to question the veracity of whatever they professed. It was clear they were slobs. Moreover, I had some weighted though tacit assurance that in *my* profession slobs did sloppy work. I steeled my tie-fortified mind to joust with them and admonish them from knighthood.

As months passed, it became clear that neither would they be banished nor were they slobs. In fact, they were often bright and careful workers. I became so comfortable that I decided rebellion might be my best course. With temple-pounding courage, I went to work one day without a tie. I walked in among my colleagues and, with shortened breath, I waited. Nothing happened. I sensed disappointment. *I wanted a reaction.* So badly did I want a reaction that in the time that followed *I wouldn't wear a tie even when I wanted to.* I loudly pronounced my abandonment of the tie, and again sensed disappointment because no one seemed to care much. At some point in time, I forgot the tie. I also forgot to hate it. And somewhere in that process, I sensed a peacefulness settle over me about ties . . . but, more importantly, about myself. I discovered I could be my true self, with or without a tie.

As a result, I have become more tie conscious, but in a remarkably different way. I am now alert to whether or not my wearing a tie would cause someone in my presence to be so discomforted that they would experience the detachment I once did. I either wear or don't wear a tie, depending how its presence or absence might affect others. My grace of being with others has now replaced convention and my anti-authoritarianism.

To restate this: at one time there was a cultural convention that said, "a professional (male) must wear a tie." That convention and thousands more like it exist in each culture on this planet. Yet currently, there is a growing awareness that each of these conventions is losing power to describe the essence of a person. The trappings of time are gradually dropping away to expose the worth of persons in an intrinsic way. Does this newly exposed, newly self-reliant person stand with the values of Salk's Epoch A or Epoch B? Do they express themselves as humans who have risked introspection and adopted substance, or are they the narcissists of darkness who use introspecton as a means for exploitation?

The fearfulness of this quandary can be attributed to these evolutionary times. Since cultural prerogatives have established and enforced the affiliation, empathy, and autonomy sequence in the past, *we have no precedent to judge the present.*

In his book *The Metaphysics of Modern Existence*, Vine Deloria senses the anxiety for this posture as he writes:

> Changes in culture and conceptions of civilization can be compared with the gradual change in the pieces of a mosaic. At a certain point in the process of replacing one piece with another, we no longer have a clear picture of what we had, and we do not yet have a clear picture of the new pattern we are creating. There is a dreadful middle ground in this process of transformation in which we simply substitute meaningless pieces over and over again. Change, even transformation of attitudes, is not an end in itself unless it leads to a more profound and comprehensive idea of the meaning of existence.[1]

Yesterday's so called "autonomous" human, the Authoritarian of my view, was the citizen of Consciousness I, the rugged individualist whose soul was self-judged worthy or flawed on the basis of strict and rigid adherence to external, doctrine-based authority.

Consciousness II and the technocratic ethic did no better in providing us with a viable portrait of the truly autonomous or intrinsic person. Instead of servitude to the "truly believed" authority of a culture, the Consciousness II or Dependency/Manipulative person was in bondage to process-based belonging and productive affiliation.

It is almost as though natural evolution has proceeded along a path of preparing humans to externalize their individual being as well as retain their individual being. The externalization processes created culture and the aggregate tendencies to define the individual in terms of the group. However, I propose this tendency has created technologies that demonstrate the limits of usefulness of this idea. Unfortunately, enthusiastic citizens of this technology retain their collective immersion and their blindness to the new epoch. Their shopworn platitudes, archaic defensiveness, and willing retreat into competitive and closed-mind systems are their only comfort. Yet, strangely, the very toys their technologies have created insure the path toward the intrinsic.

Figure XIV shows the synthesis of the major ideas to this point. Salk, Harvey, Reich, and my own work are graphically portrayed in this image.

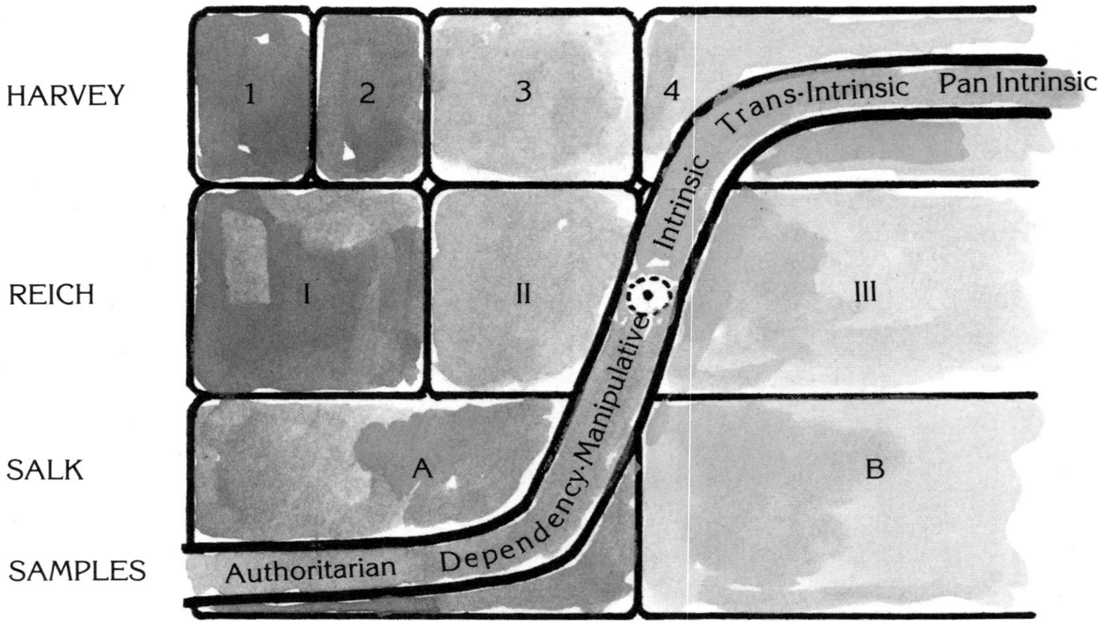

Figure XIV.

The major attribute of this synthesis portrayed in Figure XIV is that it represents the work of four people whose focus, backgrounds, and intentions were all different. Readers will no doubt be able to superimpose the work of dozens of others on this model and enrich and amend its perspectives. Also, as said in the introduction, I have not chosen to utilize the writings from esoteric traditions. These are findings from *within* my world. I am a citizen of the Western, technocratic heritage . . . I am also a child of the planet. In no way is this an intent to deny the wisdom of Taoist, Hindu, Brahman and other traditions. Rather, this is a commitment to the notion that *all* of us have *all we need to know* to invite ourselves back into harmony with the planet.

Salk's Epoch A—the Epoch of competition, reductive vision, isolation, external restraints, and self repression—is a perfect context for Reich's Consciousness I and II. Also, the personalities most likely to succeed and maintain sustenance under these conditions are Harvey's 1, 2, and later, 3.

As the inflection point is reached and achieved, a major transformation in all aspects of the interrelationships emerges. Salk sees it as the beginnings of Epoch B. Its attendant values of cooperation, proliferative vision, simultaneity, internal celebration, and self-expression begin to become adopted. Reich calls this Consciousness III, and Harvey's System 4 personality is best nurtured by such a context. For me, the evolution is characterized by the maturation of the Intrinsic ecology-individual to the Trans-intrinsic which brings a dynamic unity to all contexts of culture and person. Eventually, through the inevitable dawning of the planetary era, the Pan-intrinsic ecology-individual flourishes.

Many current writers have used the phrase "paradigm shift" to explain our toying with the inflection point of this transformation. A paradigm is a pattern of the way knowledge is used. A paradigm shift is a shift in use of the pattern.

For me, our journey through the thoughts, ideas, and the speculations of these pages is an expression of the shifting patterns of knowledge. Knowledge acquired, codified, and expressed by the culture to which we belong. As a paradigm shifts, we have historically believed that new knowledge was necessary. Yet, it is highly likely that this is not so. *It may be that only the way we use the pattern needs to shift.*

Recall back in Chapter 3 when we explored the concept of a macron . . . a pattern that emerges when material substances are in resonant harmony with the energies they experience. When energies change, patterns change. Perhaps it is a new, or more likely old, energy that technological people are sensing. Perhaps it is an expression of an innocent entrainment of the natural person within us, back into harmony with larger natural systems.

Perhaps we are responding to the process of *morphogenesis* discussed on page 57. Perhaps we are beginning to resonate with the holonomic pulsebeat of the cosmos. We are beginning to honor the notion that mind is as much a child of cosmic patterns as it is a pattern created by the brain. Perhaps the withering of closed-system being is liberating every context of our relationships . . . those with self, with others, with life, death and the universe.

Our evolutionary excursion which denied our linkage with the whole has born fruit as pithy and empty as we have become. Our consciousness is not new as much as it is re-patterned, our mind-body growing in its inclinations toward unity. Disunity—dissonance with natural harmony—has created cancers in our consciousness as surely as within our bodies. It is time to heal.

PART IV
Worlds in View

As vital as are the implications drawn from historical, biological, and psychological evolution, another even more exciting issue emerges. This emergent issue is the nature of the evolution of consciousness. Chapter 12, *Mind Evolving*, extends the premises of Chapter 1, *Mind, Beginning*. Consciousness is born of the physical stuff of the Universe, that patterned form of energy we call matter. Matter in its simplest forms possesses innate tendencies to configure with other forms. Each configuration follows in form an expression of this consciousness. The evolution of consciousness is thus not bound to the time-ordered sequence of events inherent in classical, historical explanations of social and biological evolution. At first, the process is virtual or non-reflective; later, its complexity results in reflective consciousness.

Chapter 13, *East/West: North/South*, is an exploration of how different cultures have created forms of consciousness consistent with their chosen patterns of mind function. The most obvious, and by far the most popular, distinction lies in the comparisons between Eastern and Western consciousness. These are examined briefly, and then what is considered a far more important distinction is made . . . that between Northern and Southern consciousness. Northern hemisphere consciousness includes both Eastern and Western traditions and has historically weaned consciousness away from nature and deeper into cultural vision. Southern hemisphere consciousness has invested far less in culture and instead has maintained the centrality of nature in all its institutions.

Chapter 14 *Nana I Ke Kumu. . . Look to the Source* celebrates the possibilities of blending the technocratic consciousness of the North with the more planetary vision of the South. Distinction is again made between religiosity and spirituality with spirituality cast as the primary medium through which a holonomic, evolutionary world view may be created.

12

MIND EVOLVING

Several times throughout the previous chapters, I have related the contexts of evolution and consciousness. In a way, the reader might have been confused with regard to which kinds of consciousness were being dealt with in these instances. Let us try to clear this up. In using the term *consciousness* throughout this book, I have not been specific because the holonomic model of unity and wholeness is most appropriately used in a non-reductive or non-taxonomic way.

My first assumption with regard to consciousness is that it is universal. In keeping with the Native Indian traditions with which I am familiar, I am comfortable with the idea that all things, living and non-living, have consciousness. Recall in Chapter 1 that I alluded to the work of Erich Jantsch and his honoring of virtual and reflective consciousness.[1] Jantsch traces this specifically with regard to an ontogenetic model of the evolution of human consciousness. Ontogeny is the way the individual develops from sperm/egg consciousness to mature human wisdom. It is the path that leads from aggregates of biological molecules to *homo sapiens*. It is a journey that includes all the consciousness of all the creatures that *sapiens* once lived through. These are the songs of consciousness, songs of the species, sung through the individual, echoing through billions of years of planetary time.

Jantsch does not specifically sponsor the extension of virtual consciousness to all material in the universe, yet his work does not exclude it either. Therefore, I will specify an extension. *Virtual consciousness, I then suggest, is that form of "consciousness" expressed by the response of the patterned energy we call matter to other patterns in a holonomic universe.* Electromagnetic, strong/weak, nuclear, and gravitational forces provide the configuring patterns to which matter responds.

Thus, with regard to the overall context of consciousness and its universality, everything possesses consciousness. The harmony of the consciousness of matter is expressed in the regularity of the periodic chart of the elements and the predictability of the world of physical and general chemistry. In the methodologies of contemporary Western science and technology, *virtual consciousness* is voraciously utilized by tens of thousands of industries, laboratories, hospitals, and manufacturers. We make alloys, pesticides, non-spoiling milk, wristwatches, cloud seeders, lipstick, and scotch whiskey. We chemically alter bloodstreams, brain chemistry, estrogen production, and reduce acne. We recombine DNA, genetically engineer and test one for rabies. But I have yet to hear any common and unqualified expression on the part of a scientist or engineer that matter has consciousness. All will say matter (elements) or aggregates of matter (minerals) have properties, but I have heard none say consciousness.

So for me, virtual consciousness is the most basic form of consciousness that humans can know. By basic, I do not mean base in the sense of trivial, unimportant, or vile. But rather it is basic in the sense that it is the form of consciousness expressed in atoms and in the cosmos.

The Western world view is shattered by perceptions that extend toward both the infinitely small and the infinitely large. Our consciousness has traditionally been defined by our intentional *separation* from the atom and the cosmos into a Newtonian realm of the world of "our size." As our perceptions extend to the world of the atom and its patterns, our conventions disintegrate. Similarly, if our perceptions extend toward the boundaries of the universe, then there too our minds sever the umbilicus of Newton's realities.

Let's look toward the harmonizing of pattern into expressions of consistency in the non-consistent world of sub-atomic physics and atomic physics. When we postulate a consciousness for matter, the possibility for a higher state of unity exists. I can tease my mind, my being, my own personal consciousness toward higher holism. I become, intentionally, more a part of, rather than separate from.

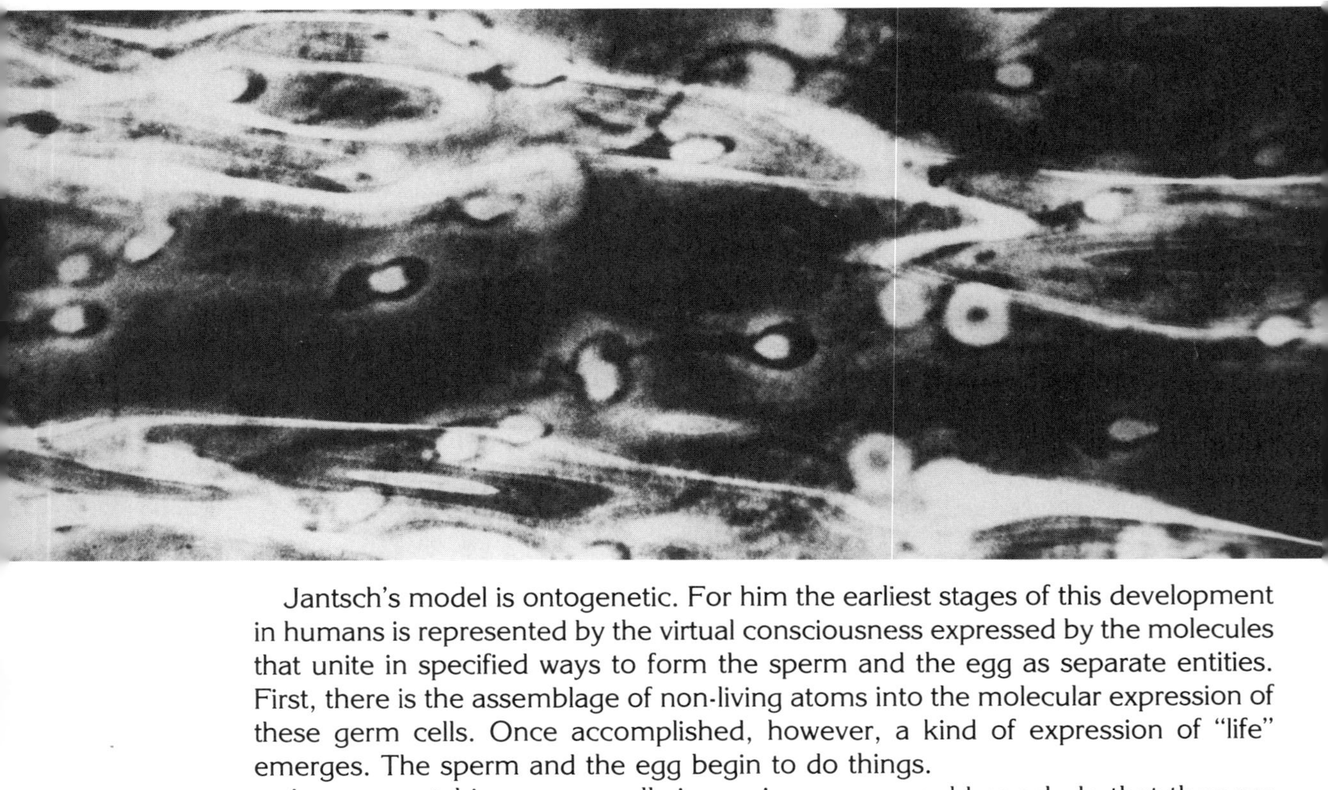

Jantsch's model is ontogenetic. For him the earliest stages of this development in humans is represented by the virtual consciousness expressed by the molecules that unite in specified ways to form the sperm and the egg as separate entities. First, there is the assemblage of non-living atoms into the molecular expression of these germ cells. Once accomplished, however, a kind of expression of "life" emerges. The sperm and the egg begin to do things.

Anyone watching sperm cells in a microscope would conclude that they are alive. Because of our knowledge that they come from within a male human, there is still a reticence to defend that view. If, however, we found these little whipping-tailed ovoids swimming in pond water, we might instantly call them alive and consider them an organism.

But a sperm is too incomplete to be considered "alive" unless it unites with an egg. The watched sperm cell remains active for a few days, and then begins to disintegrate and be resorbed by support systems. It "dies," so to speak. But wait. Isn't this the point? If it dies, was it not alive? Yet, it is clear that sperms do not make other sperms. Something else makes sperm cells. However, they do look alive, and they seem to be doing things if you watch them. Perhaps they *are* alive, and perhaps they *do* possess consciousness.

If an egg cell is inserted in the medium in which the sperm cell exists, then the sperm cells begin to swarm to the egg. If conditions are right, then something quite unique takes place. The sperm fertilizes the egg. More appropriately, they fertilize each other. Two forms of *virtual consciousness* unite to create the con-

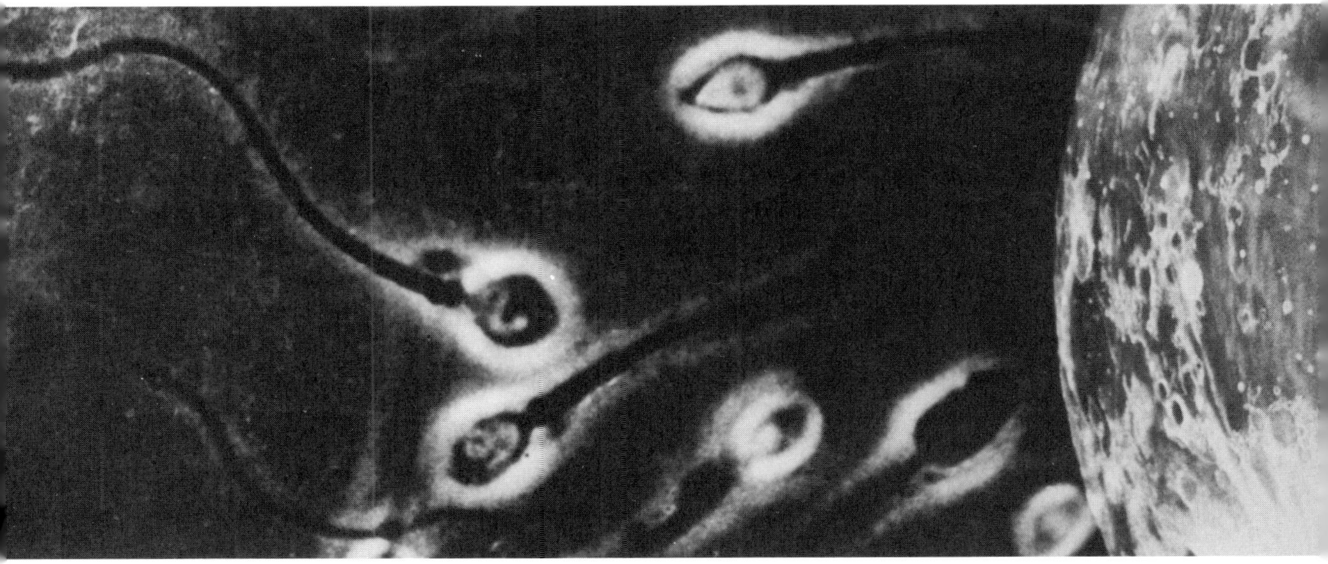

text of a "new" consciousness. Neither the sperm nor the egg could do alone what they can now do together. Both seem to have "lost" their original identity, and new patterns emerge. The new consciousness allows the egg cell to divide and, as it does, it invites virtual patterns from a sustaining nutrient "ocean" that bathes it continuously.

The new consciousness is perinatal. As the cell continues to divide, selecting with a grace and harmony that has been danced by all matter and sung by all energy, life as we choose to define it begins.

In the holonomic model, the shifting nuances of this process continue to be an expression of the resonating harmony that patterned energy must express. What is happening is holonomic; it is *holos* . . . it is holy!

But now a new context of consciousness emerges. The growing fetus, while establishing unity with the mother, also establishes separateness. Perhaps the first concept of "otherness" is experienced. Until this time, consciousness has been designated primarily in the virtual or non-reflective modes. Although contemporary arguments abound about when there is fetal consciousness, I will assume it is coincident with conception. Certainly, when the mother can feel movement of any sort within her body, it is likely the fetus has felt movement within his or her environment. It *knows* it is not alone, because *a movement has happened to it!*

There is little doubt in the minds of most that there is consciousness in utero, and there is the distinct likelihood that it is in utero that much of social consciousness is conceived. Studies have shown that infants learn the music of their mother's language in the oceanic utero and learn the words in the atmosphere with their culture. However, the "social" consciousness is experienced through another. Thus it represents the most fundamental unit of social presence.

Tracing the unborn child's progress to this point, it is easily seen that up to this quality of consciousness, nature has been the primary affiliation. Unity has been the norm, and this first expression of separateness is the newest awareness. In effect, the only trend away from a consciousness of unity was the discovery of the "other" (the egg in the sperm's case, and the sperm in the egg's case), just prior to fertilization. Once fertilization was accomplished, unity was again restored. A different "other", however, is now imposing its presence on the child.

Labor begins. The "other" is now unavoidable. A whole world of events that the child is nearly unable to control is now taking place. It is the second level of consciousness. Historically and traditionally, this has been the event of most exaggerated importance. The trauma of birth.

In Western society, the achievement of the quality of consciousness associated with the emergence from "oceanic" knowing to atmospheric consciousness has been the most feared of and interfered with context of human life. Western men and women had moved so far from nature that contexts of pain, sin, death, and responsibility shrouded the aura of birth. If the mother reflected the social aegis through fear, it is more than likely that the child was fearful too. Society insured that birth became hell. So hell-filled that Western medicine transformed it into the likeness of a medieval torture chamber. Needles, drugs, anesthesia, knives, scissors, stirrups, masked attendants, and the whole lot were assembled to reinforce the unnaturalness of birth. Birth had been wrenched from nature and armored with the medical weaponry of culture.

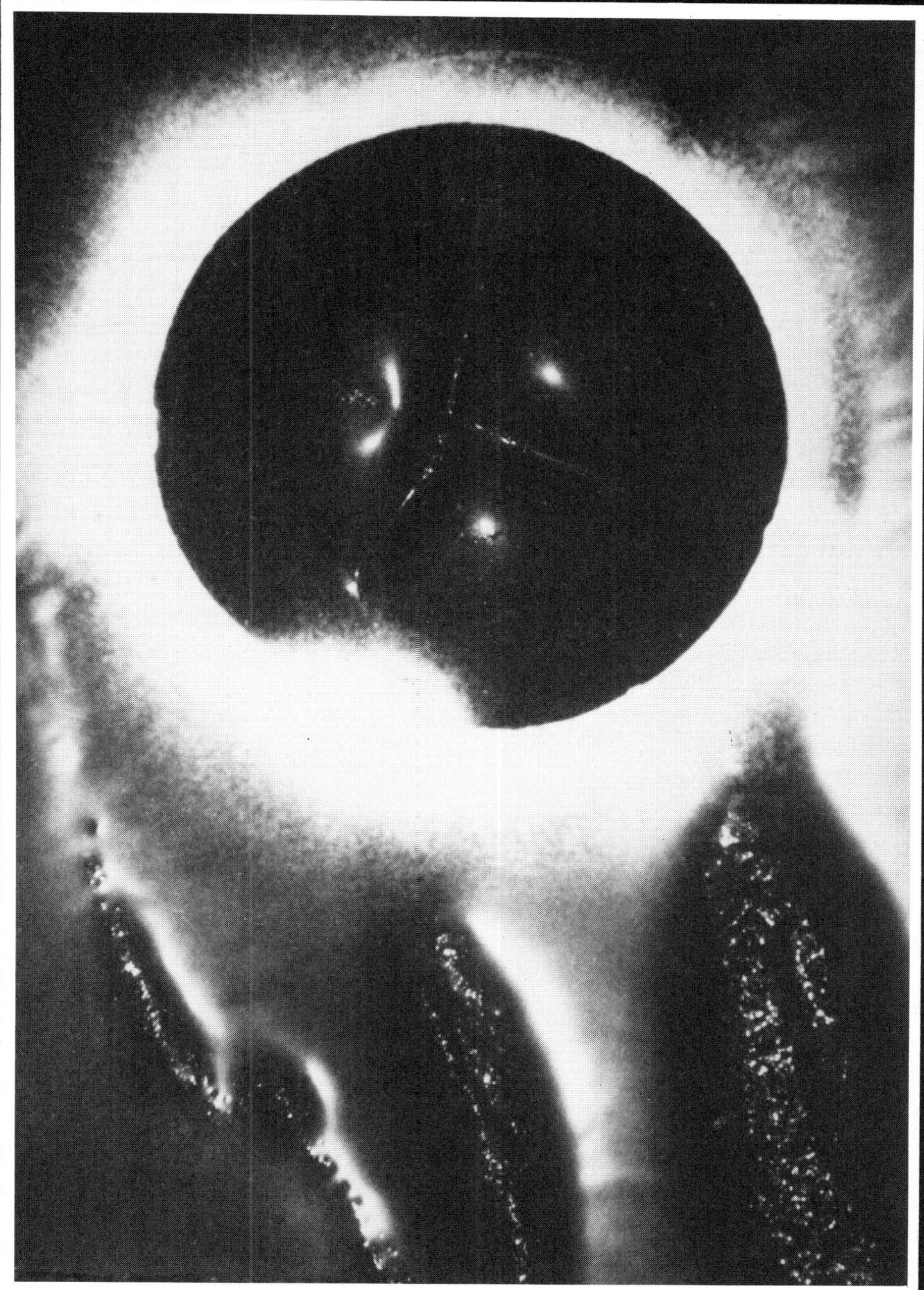

The transformation of consciousness from oceanic to atmospheric unity was almost completely altered by this level of cultural intervention. A fearful mother and a fearful child lost synchrony. The child lost the capability to resonate with the mother's rhythms, and the mother lost the rhythm of nature. Birth trauma clearly was a reality. But it was less a reality of nature than a contrivance of culture.

Today, the horrendous intervention of culture in the birth process is a choice rather than a mandate. A hospital or community of any size will generally offer the option of knowing more fully about natural birth. This is the natural versus technological. Medicine is not abandoned in this emergence, but rather restored to its role of servant to culture and nature instead of tyrant.

The born child is in the emergent atmospheric and non-oceanic cosmos. It is born into free space. From this day onward, consciousness will no longer be physically bound to the sensory world of the oceanic environment. Instead, the child's forms of knowing, learning, being, and transforming become atmospherically free. But is this freedom complete?

The newborn child needs help. As Jerome Bruner once said, human gestation lasts until the age of sixteen. But Bruner's comment was admittedly conservative and exaggerative. The human infant does take longer to become quasi-independent than the newborn of other species. A year to walk, three to use much conventional language, and seven or eight to be reasonably self-caring (although not self-sufficient). Bruner's description was influenced by culture. Current law requires school until sixteen. The law serves the wishes of the culture to insure that the nature of the newborn's infinitely free consciousness be appropriately sculptured by the society. From birth to sixteen, our culture guarantees the processes of inhibition of consciousness.

During this period, two more of Jantsch's consciousness realms are usually experienced. From the newborn's ecstasy in finding free space, the child moves to an awareness of other free-space beings. The child sees society, and begins the process of establishing relationships in the presence of the free space of others. To do so, the child learns that the physical confinement of the womb has been replaced by the psychological encapsulation necessary to function in a world filled with others.

The best formulae for this getting along are written into the rules of the culture. . . the folkways, customs, and mores. So those of free-space consciousness adopt the historical psychological pathways of the culture into which they were born. Eventually, if they do this well, they are afforded access to a limited kind of "free space" again. They can direct their own action.

On the way to this new consciousness, they have experienced the process of conformity, been recipient of the control of others, and learned how to survive via coping. If they don't do this, they become frozen at the closed-space level of establishing relationships. Until recently, many in our society were fixed at this level. Along the way, our newborn has grown into an adult. In so doing, an ego has been established. Ego is that self-made description of the essence of one's being that each person uses as their court of last resorts. It is their personally-built sanctuary.

The new free-space consciousness, that is, the level of directing one's own action, is strongly influenced by ego. Ego supports the action. In a sense, the nature of the ego is not as important as the fact that a person had the strength to break from the pack. The previous stage, the relationship stage, is characterized by pack consciousness. This new consciousness is characterized by a limited kind of autonomy, the kind I earlier called Intrinsic.

However, the achievement of this new consciousness may also be viewed as separatist. But its separation is primarily cultural. Most who would achieve this stage would recognize a distance from unity, a distance from the natural. The next quality of consciousness would be a return to unity via a transcendence of ego or personal boundary consciousness to a sense of unity with all things. The world of humans—the world of nature—seems to make sense without specific understandings. Jantsch uses the phrase "acausal orderness," suggesting one can sense order without discerning the reasons.

Once this is achieved, the next cycle of transformation involves the application of this new consciousness to more inclusive and more holistic contexts —ecology, synchrony between human and natural systems, and, finally, a transcendence of the separateness between humankind and the planet/cosmic systems.

Jantsch's system eloquently synthesizes the work of Grof, Lazlo, Franz, and Jung, all of whom felt that the consciousness of unity was the innate tendency of humankind. Many in the past have stopped in their progression up this spiral and sought completion in terms of the level that they had achieved. They argued only on the basis of the level at which they rested and the ones that preceded them. Such narrowness is cultural and not natural.

The approach to consciousness in this book is clear. One *must* experience all of consciousness, whether or not they choose to. However, one may also choose to ignore all but a narrow realm of the whole spectrum.

Figure XV.

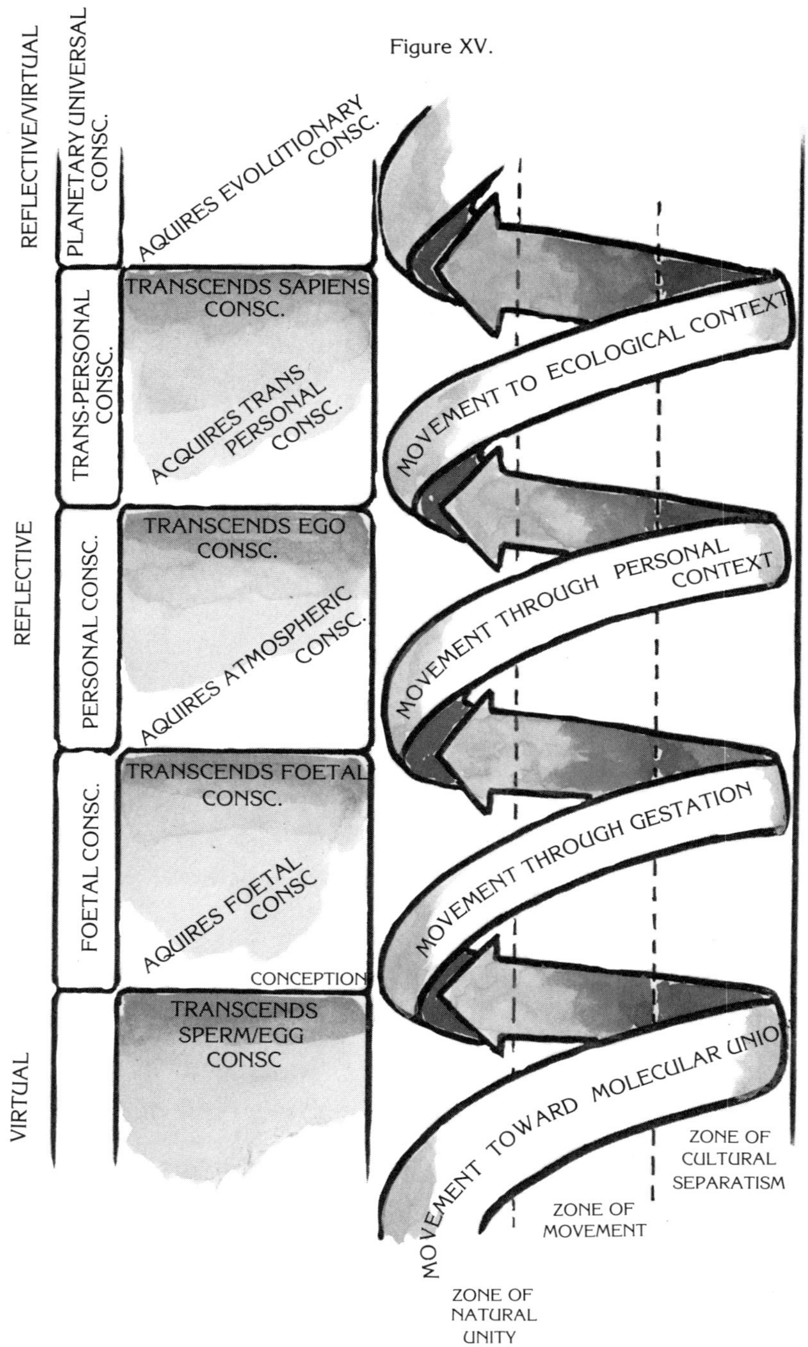

REFLECTIVE/VIRTUAL
PLANETARY UNIVERSAL CONSC.
AQUIRES EVOLUTIONARY CONSC.
TRANS-PERSONAL CONSC.
TRANSCENDS SAPIENS CONSC.
ACQUIRES TRANS PERSONAL CONSC.
REFLECTIVE
PERSONAL CONSC.
TRANSCENDS EGO CONSC.
AQUIRES ATMOSPHERIC CONSC.
FOETAL CONSC.
TRANSCENDS FOETAL CONSC.
AQUIRES FOETAL CONSC
CONCEPTION
VIRTUAL
TRANSCENDS SPERM/EGG CONSC
MOVEMENT TO ECOLOGICAL CONTEXT
MOVEMENT THROUGH PERSONAL CONTEXT
MOVEMENT THROUGH GESTATION
MOVEMENT TOWARD MOLECULAR UNIO
ZONE OF CULTURAL SEPARATISM
ZONE OF MOVEMENT
ZONE OF NATURAL UNITY

Yellow light seems discrete from green, depending on how you look. In terms of color, they are separate only if you look at them separately. If they are looked at in the visible light spectrum, then one can only separate them arbitrarily. There *are* clearly tendencies toward greenness, just as there are tendencies toward yellowness.

So too it is with culture and nature. Our consciousness can trend away from nature and focus upon cultural elements. But we cannot stay with culture too long before the natural impinges again upon us. At that time, nature beckons us back. Each time we return, a new unity is achieved. The process is summarized in Figure XV.

A tracing of this process in Figure XV reveals a kind of resonance between nature (unity) and culture (separatism). The shifting back and forth as the evolution of the individual human being takes place is an extended indication of the well-known axiom of embryology, *ontogeny recapitulates phylogeny*. The meaning is that ontogeny, the growth of the fertilized egg—through fetal stages and eventually to a full-term child and mature human—reflects the evolutionary path of the species. It means that each of us as individuals in our mother's womb revisited the various forms that our species experienced as they journeyed through the sweep of evolution. I argued in a previous book, *The Metaphoric Mind*, that this premise allows us to postulate that—in terms of virtual consciousness—we effectively know what all things knew.[2]

We are, as far as consciousness is concerned, first a child of matter, then of one called life, then of all the multi-celled creatures that gave their form to our being, and finally a child of *homo sapiens* with our mother also a *sapiens*. So goes the natural sequence. Now comes the cultural one. In our *sapiens* stage with our mother, we begin to recognize the presence of culture. Our birthing assures it. Then comes family, school, and all the other institutions. Relationships, rules, and rituals—how far away it all now seems. The amnionic ocean becomes a swimming pool, all the things I once was are now pets and residents in zoos . . . or, more likely, pickled on the shelves of academia. My attachments to matter and the ghost breath of the stars are replaced by Boy Scouts and vesper services.

For most, planetary consciousness—that is, the emergence of a consciousness of evolution—is not life's primary concern. Yet, strangely, it may well be the tacit concern of the evolutionary process itself. This ontogenetic model of evolving consciousness in the individual human may be the map for the species. If, as Salk contends, survival is encoded within us, then this may be but an ontogenetic statement of a phylogenetic tendency. Perhaps it is the possibility of the emerging consciousness that is the true attribute of genetic encoding.

I believe it is fair to say that most humans in the United States are more concerned about how they relate to others, than how they feel about themselves. After all, hasn't how they feel about themselves been vilified as narcissistic? Yet is is clear that just as evolution takes place at the level of the individual and *not* the species, modern humans are *beginning to respond to the evolutionary urges of their own consciousness*. This is neither a cultural, political, nor organizational tendency; it is an *individual* one.

Against all of this is a tapestry of contemporary crisis. Money, material, standard of living, diet, and health exist against a changing sea of inflation, unemployment, and changing expectations. We are truly in crisis times. Most are too busy to question whether the crisis lies in our basic values, or simply lies in the times.

Daniel Yankelovich in his insightful book *New Rules*, cites a basic shifting in values in contemporary American society. He traces the post World War II dream of affluence and an eternal "good life" as it crashes headlong into the economic and resource constriction of the early 1970 s. He, like Christopher Lasch, sees a crisis in the basic fiber of American society. However Yankelovich recognizes that resources are diminishing, rather than expectations. Lasch took the struggle for self-fulfillment to be an indication of the decay of integrity in the American character. Unlike Lasch, Yankelovich saw the quest for self-fulfillment for what it really is . . . a search for a new philosophy of American life. Affluence has given way to hard times and still the quest goes on.

Americans do not yet have a language for what is actually happening. However they do have a vision. The vision is subtle, confused and diverse. Even pollsters like Yankelovich, who usually hear reflexive mutterings rooted in the past as they question the citizen on the street, are sensing an emergent quality of expression. It is growingly clear that we are *not* witnessing a retreat into the past. Only those who mistakenly see the search for self-fulfillment as a return to the Authoritarian version of self reliance of Consciousness I, are vulnerable to such interpretations. We are in reality experiencing the early phases of a cultural evolution in which we are somewhat grudgingly and hesitantly giving up material dreams for dreams of holonomy. We are trading dreams of consumption and exploitation for dreams of harmony and blending.

In the opening words of the preface to his book, *The Promise of the Coming Dark Age*, Leften Stavrianos says:

> This is an exhilarating moment in human history. Far from being an Age of Trouble, or Alienation, or Anxiety, this is rather the age in which man for the first time can turn from his age-old struggle for survival to a new struggle for self-fulfillment that will enable him to realize his humanity. *Homo sapiens* is on the way to becoming *homo humanus*.[4]

Homo humanus!

What could Stavrianos mean? He could be implying exactly the same thing this book is about. We stand clearly as a culture on the consciousness threshold between the personal and trans-personal levels of Figure XV. We stand on the bewildering threshold of Reich's Consciousness III. We stand cautiously in the hedgerows of Salk's inflection point.

The bewilderment, caution, and timidity we are finding come to us from the past . . . not the natural past, but the cultural one. No wonder we cannot entertain transformation and evolutionary consciousness in the light of our cultural training. But does that give us permission to stagnate . . . to give *more* of our personal power to the archaic prisons of our cultural history?

I think not. The beckoning is becoming too loud, too persistent. It drifts out of the disillusionment of the past. The dark age may be nearly over.

13

EAST/WEST NORTH/SOUTH

During the past decade, few things have caught on as quickly and as passionately as the "discovery" by large numbers of Westerners of so-called Eastern consciousness. The ancient wisdoms of China, Tibet, India, and Japan have been the most popular and generous sources. Relatively recently, the ancient wisdom of the non-Judeo-Christian mid-east has become sought after through the penetrating humor of Sufi tales and parables. Alan Watts, Idries Shah, Maharishi Mahesh Yogi, Ram Dass, Sri Auribindo, and Gurdjief have webbed the nation with their writings and presence at overflow crowds.

The popularity of Eastern consciousness surprises many and worries some. Many who view the Eastern teachings as religion and doctrine fear its encroachment on the basic faith contexts of Western Judeo-Christian domain. To them its presence is Anti-authoritarian. The fears of the traditional ministry are well-founded for a surprising number of young Americans are disenchanted with traditional Western religion and are flocking to the Eastern teachings as if they *are* religion. Even though in the strictest sense Buddhism, Hinduism, Brahmanism, Taoism, and Sufism are *not* religions as Western humans know religion. Yet these philosophical world views are treated as religions by zealous Western converts, who are quick to replace old zealotries with new ones.

But apart from the religious context, Eastern wisdom is popular in the West as a philosophical expression of world view. There is far more holism inherent in all levels of Eastern thought than in Western thought. Alan Watts will probably remain the most cogent Western synthesizer of the Tao. Tao (pronounced dow) Watts described as the "watercourse way." He then goes on to create the notion that "the Tao that is the true Tao cannot be spoken of."

Further, he weans the Western mind from its reflexive quests for certainty as it paranoically demands reductive exactness. To try to reduce the Tao to exactness destroys our understanding of the Tao.

Our only contemporary counterparts come from sub-atomic physics and the dawning of awareness that at the level of most minute certainty, certainty no longer exists. Fritjof Capra, Itzhak Bentov, and Gary Zukav are most eloquent spokespersons for this. Because sub-atomic physics and its other scale counterpart, cosmology, are not commonplace arenas of thought for the masses of Americans, the enigmatic riddlery of Eastern thought is more approachable.

Eastern thought's most vital gift is its introduction of substantive ambiguity to Western thought. Substantive, in that there is an intellectual methodology to the thought, and the thought contains a built-in intent to create an acceptance of non-reductive, yet tentative, holism. Consider the following:[1]

> The Tao that can be told is not the eternal Tao.
> The name that can be named is not the eternal name.
> The nameless is the beginning of heaven and earth.
> The named is the mother of ten thousand things.

Such ambiguity and convolution in reductive thought could never survive the first reading in any Western college or university theses. Yet, these words and their like provide the substantive world view for nearly half the literate humans on this planet.

It is likely that there exists in the conflicting methodologies a way of creating a world view, East and West, that may be rooted in the psychological perception of the individual in the cultures involved. Jean Shinoda Bolen addressed this issue in a marvelously readable book called *The Tao of Psychology.*[2] She suggests that C.G. Jung's idea of synchronicity lies at the Western heart of understanding Eastern wisdom.

Shinoda Bolen traces a vital connection that relates to my earlier distinctions between unity of the individual's psyche and the cultural ecology. She comments on page 3 that in modern translations of the New Testament into Chinese, logos is translated as Tao. She further cites that the gospel of St. John now opens, "In the beginning was the Tao." Following up on that distinction, it can be concluded that *logos*, (cosmic reason of the Greeks) quickly became *-logy* in Western thought. Logy . . . the science or study of. Once linked with science, Western *logos* became reductive in the patterns of logic and analysis. To those who would create the adversity between science and church, a theological connotation of *logos* persisted. For the clergy, *logos* was The Creative Word of God, identified with cosmic reason. Though the religious word *logos* was separated from the scientific *logos*, the reductive methodology of *-logy* remained the same. Theological journeys, from the origins of *logos*, were as reductive as the path chosen by science.

Jung's idea of synchronicity struck many of his contemporaries as non-scientific heresy. After all, the Newtonian, Cartesian visions were imbued with logic, sequence, and order. For events to occur in the world separate from the reductive certainty of order was folly. It would presume a larger pattern than could be accounted for by science, a pattern which defied definition and could not be reduced to verifiable presence. Coincidence could easily explain sychronicity. Coincidence was random. Once coincidence was recognized, it could easily be linked to some logical causality. If not linear, it would be expressed by the logic of probability.

Synchronicity is *acausal*. Cause and effect provided the litany for the Western mind in science *and* in religion. Newton gave numbers and value to opposites, perhaps without the realization that Western theology and philosophy was supported from its very core with dichotomy. Paradoxes, dyads, dualities, opposites, black-white, good-evil, and hundreds of paired concepts burned into the neural network of Western thinkers. One of anything became an incomplete reflex-arc in the Western mind.

No wonder the Tao and Hinduism so excite the Western mind. They offer it openness. Recall the closed-system mind discussions of Chapters 5 and 6. Such ideas counteract the dominant experience of an individual living in the ecology created by our Western technocratic culture. The quest for certainty and closure is the most powerful ally of closed-system thinking. For us, certainty lies in closure. For the Eastern mind, certainty lies in the preservation of openness. The *logos* nurtures closed-system thought, while the Tao nurtures open-system thought.

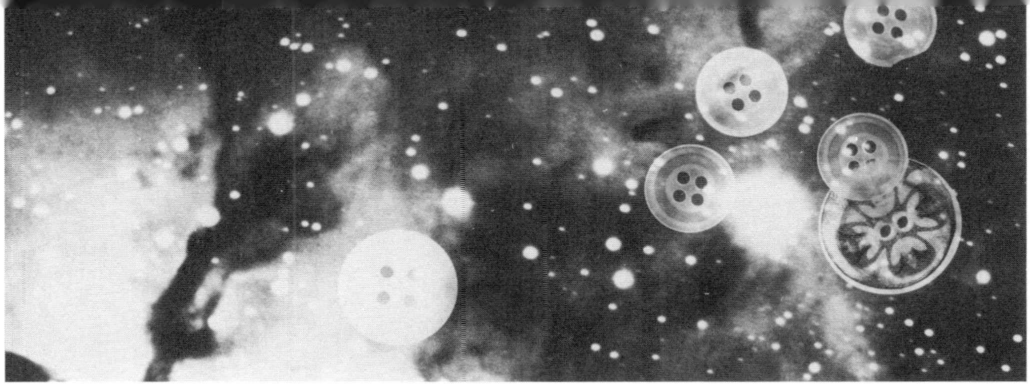

For example, in Western science, light or radiation—like time, space, and matter—have been central unifying concepts. Western science reveled in the discovery of the telescope, the prism, and spectrometer, and any number of devices that made light more readily categorized and tucked into closed-system cubicles. With such devices, the universe was mapped. We learned how to fix light and its sources. The Atlas of the Universe was published with prestigious fanfare.

But then, an embarrassment. Someone began to map the places where there was an *absence* of radiation. These empty places too were mappable. Remember what Fritjof Capra has said about emptiness: "this emptiness . . . is . . . the essence of all forms."[3] This somehow was anticipated in the Sufi story of the Mullah Nasrudin who once was found searching for some lost buttons. "Mullah, what are you doing?" a friend asked. "I search for my lost buttons," replied the Mullah with taxed patience. "Where did you lose them?" his friend persisted. "Over there," gestured the Mullah to a place in the darkness. Bewildered, the Mullah's friend looked toward the shadows and then asked, "Well, why do you search for them here?" The Mullah raised from his knees and said tersely, "The light is so much better."

Western scientists never found the black holes in space until they mapped the darkness, until they looked where the light was *not* better. *Logos* is a limitation as sure as it is an advantage.

It should be clear to all who would witness these differences that the Eastern thought context has nurtured an exclusivity of its own . . . and that would be a tacit exclusion of closed-system thought. Yet, make no mistake. It is more difficult for the habitual, closed-system thinker to become open than for the open-system thinker to accept temporary closure.

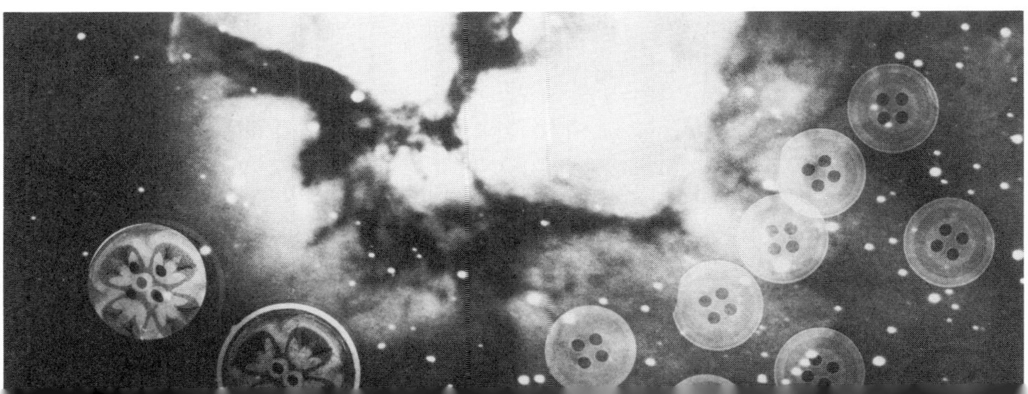

As we look at the worlds of East and West, we can detect the hybridization of both. In the West, a murmur ripples through the "new consciousness" community when Ram Dass is scheduled to speak. In Asia, a banter of excitement accompanies the announcement of an IBM seminar on advanced applications of computers to industry. The West, resting on the already rusting battlegrounds of technology, seeks redemption from its self-inflicted imprisonment of mind. The East seeks equality as it imports the time-lock cells that closed-system Western thought have perfected. Both see progress as developing an equilibrium of thought at the expense of the planet. They are less and less East and West as they become more North.

A chance conversation with Jean Houston—that vital, brilliant, high priestess of the transformation—gave birth to an idea that had been gestating in me for a long time. I remarked to her that I didn't think East and West was as important as North and South. Her attention riveted. She asked me to say more. I did. I claimed that nearly all the mythic traditions of the North were myths of intervention and contained the intent to dominate nature. The South, I argued, had its mythic tradition focused on harmony with nature. Their idea of "winning" was not in conquest of nature, but rather in living *with* nature.

Jean was a long-time colleague and close friend of Margaret Mead. Almost instantly, she began to flow with insights in regard to this topic. She and I talked well into the night. It was clear to us that there was an incredible difference between the Northern and Southern hemispheres. Further, we concluded that it was best expressed in consciousness. Following the creation of an ecology of consciousness, any culture would evolve its institutions to best represent equilibrium with that consciousness. These initial conversations with Jean were vital to the following two years of my search to make sense of this. Jean generously contributed to what follows.

The consciousness of the North was manifested by two contexts of consciousness that utilized nature as a resource rather than as an ally. The West blatantly exploits nature as a physical resource. "Go ye and populate the wilderness," René Dubos cites in his book, *The Wooing of Earth*. "The word *wilderness* occurs approximately three hundred times in the Bible, and all its meanings are derogatory."[4]

Easterners, in spite of their apparently more holistic vision, devastated the environments in which they lived. Four thousands of years before the Taoists, China was denuded, eroded and overpopulated. But, more subtly, many of the Eastern vision of nature *found in it the justification for the world views of humans*. In other words, nature became the matrix against which human modes of thought were rationalized. For example, perceived cruelty and injustice in nature were taken to be justification for it to exist in culture. The postures are basically circular anthropomorphism.

Neither of these *uses* of nature in the North did anything to prepare these human world views for other than eventual exploitation. As I have pursued an understanding of what I term the "Southern world view," I am struck by the tacit differences in how most cultures of the Southern hemisphere "use" nature. As a point of clarification, I speak of the original and residual native cultures of the South . . . not the recent European or Asian immigrants. And clearly, I do not mean the effects of colonialism. It is far easier to explain the fascism and genocide of Idi Amin in terms of European precedents than it is to explain it in black African precedents.

I consider the Native Indian people of the North American continents to be of Southern hemisphere consciousness. The tribal lands of the Native Indians are a geographic "peninsula" of consciousness that extends into the Northern hemisphere.

Let's return to the "use" of nature by the Southern context of wisdom. Nearly everywhere one searches in the tribal-based cultures of the Southern and Northern hemispheres, there is evidence of harmonious equilibrium. Since most of the cultures of this consciousness had their origins in hunting and gathering as opposed to agriculture, a higher continuity to principles of harmonious equilibrium persists through time. Even when many of these cultures moved to low-technology agriculture, the bond with natural systems was central to the institutions that they created. The Northern hemisphere traditions clearly created their original institutions from the same kind of resonance with nature that still persists in the South. But, recall the role of *-logos*. Science and religion both trying to own the description of reality drifted quickly away from nature. The benevolent presence of their originally honored natural intuition rapidly gave way to logical reductive, cultural explanations of reality. Following their dichotomistic vision, the world view of the North rapidly infused the differences between the religions and the material, heaven and earth, the godly and the un-godly.

This did not happen to the Southern consciousness. Vine Deloria says it with grace.

> Primitive people do not differentiate their world of experience into two realms that oppose or compliment each other. They seem to maintain a consistent understanding of the unity of all experience.[5]

Deloria further goes on to cite the non-recognition of cause and effect relationships. Cause and effect create within us a base of expectation as we confront the natural world. Deloria quotes Jung:

> . . . In the primitive man . . . the psychic and the objective coalesce . . . In the face of something extraordinary, it is not he who is astonished, but rather the thing that is astonishing.[6]

Northern religion, moving as it does from *logos*, creates a high degree of separation between the individual and nature. Once God interpreted the cosmos through *logos*, the cosmos became secondary. Humans separated from nature inevitably seek affiliation in something. The most available something is culture. Thus, the Northern context of religion was premised upon a separation of humans from nature; it gave humans power over nature and invited them into a power-based separatism not only from nature but from each other as well. Southern contexts of religion seem to have intuitively honored the *morphogenetic* or holonomic resonance. They seem to have retained a bonding with the cosmos by wedding their world views to *morphos* rather than *logos*.

In Southern-vision tribal contexts, the unity with nature was paramount; thus separatism became an expression of deviance. Bonding to the planet and all its forms represented far less an expression of superstition and primitive belief than it was an expression of the most poignant spirituality.

Arnold Toynbee was among the few Western historians who recognized this. He maintained that the worship of nature *unites* people because it is not "self-centered." I would substitute the phrase "self-separating" here. *The honoring of nature invites us toward the most effective transcending of ego that humans can experience.* Ego is preserved and sustained in contexts where there is *no* uniqueness. If there is *no* uniqueness and beliefs are homogenized through a religion of separatism, then ego *must* be preserved as a power of last resort.

Nature is the ultimate open system. In a holonomic view, nature represents the original harmonic of mind. But as surely as humans have origins, so too can new music replace the old. The consciousness of the Northern hemisphere expresses most clearly the way cultures can create music that is in disharmony with the origins. In Southern hemisphere consciousness, we find the ways in which cultures can stay close to harmony with the origins.

Strangely, it is one of the gifts of *logos* that has kept us from this more spiritual vision of unity. The Northern consciousness invented writing. The Southern consciousness did not. Oral traditions keep us close to our origins because each experience possesses its own unity. To share such a realization of unity, one must first experience something directly; when it is to be shared, the person must share it first hand.

The absence of written language keeps experience direct. The hearing of a story of myth from the teller of stories involves more of the human being than the reading of the same story in abstract, reductive codes. The experience of abstract, codified languages—so common to all Northern hemisphere cultures—is a means by which direct experience can be eliminated. The systematic deprivation of direct and original experience creates a wider and wider chasm between a human and the planet. As the chasm grows, so too does the awesome yearning for return.

One way the Northern consciousness has tried to restore itself is, strangely enough, through further abstraction. In 1932, John G. Niehardt published a book entitled *Black Elk Speaks*. Black Elk, a warrior medicine man of the Ogalala Sioux, became one of the first spokespersons to popularize for Americans the wisdom of the Native Indian vision. Following Black Elk came hundreds of books written about the vision of this nature-bonded consciousness. Perhaps the most popular and controversial was the don Juan series, as written by Carlos Castanedas. In addition to these sources, we are finally beginning to understand what Margaret Mead really said in her classic studies.

The technological citizen is yearning for the presence of the origins; the mother is so great that there is a systematic and intentional effort underway to re-establish pathways back into harmonies we have ignored. We may be witnessing the time of that restoration. Just as Alan Watts transformed Zen and the Tao into contexts of integrity for Western consciousness, so too do a myriad of sources invite this context of North and South back into unity.

The derogatory term "romantic" has been used to describe the traitorous Northern hemisphere citizen who subscribes to the Southern vision. "Romantic!" is the anguished cry of urbane denial of the souls' ultimate bond to the natural. This is the cry of the irretrievable technocratic human whose incest has encapsulated them within the myopic prison they call "civilized" sophistication. They have retreated so far from the *mother* that the only meaning in life is in judging wines while denying the grape. They savor cheeses but shudder at the mention of dung spattered cows. For them, the air is a list of statistics about temperature and particulate solids. The earth is pre-stressed concrete to be experienced while waiting for a taxi, and trees are noise-filled and shedding nuisances. Life is a newspaper, a bankbook, paintings on canvas; and wind appears only through French horns and flutes. Reality is what the mind of "man" has made.

Yet this citizen of technology is the true "romantic," for this person has come to love and find comfort only in the facade. Ralph Waldo Emerson as quoted by Bill Bridges, disrobed this citizen years ago:

> His relation to nature, his power over it, is through the understanding, as by manure; the economic use of fire, wind, water, and the mariner's needle; steam, coal, chemical agriculture; the repairs of the human body by the dentist and the surgeon. This is such a resumption of power as if a banished king should buy his territories inch by inch, instead of vaulting at once into his throne.[7]

But in the way the world works, this citizen is the newest of endangered species. It is this person who provides the raw material for the transformation that will rejoin us all with the flow of evolution. It is the planet that leads us now.

14

NANA I KE KUMU

. . . LOOK TO THE SOURCE

These words in the Hawaiian language serve as a beginning and as a summary for this final chapter. They offer us virtual matter, germ plasm, conception, birth, life, death, and virtual matter again. At each level of human being, we are eternally graced to be the fabric of nature. Our being is woven into all being.

Like water droplets in clouds. One time we are autonomous and float independently, and then the breathing envelope of air wafts inward and we coalesce, joining our energies with millions of others. If the inward breath continues, we become unable to be borne by the air . . . we cluster, protect ourselves in the mutual cloak of a common surface . . . and, holding ourselves tightly, we fall. To earth.

To earth, but not to death. We lose our hold on each other and find our way into the earth. There we coat tiny particles of soil and eventually find our way to the fibered tendrils of roots. Some will enter and be drawn upward to nourish the green growth that graces and softens the planet's face. Others will be swept into deeper darkness and move in ways that birth springs, geysers, and boiling, crushing torrents of mineral-rich tracings within the cores of mountain ranges yet to come.

Nothing is lost in the cycles of the mother. The earth is infinitely renewable. Its twenty billion years have seen vast deposits of what cultures have called resources be destroyed and born again in countless waves of renewal. If what we know about the universe is an expression of this infinite cycle . . . then the entire cosmos too may well be self-renewing. All the dying stars—all the planets and galaxies drifting toward darkness—possibly face the inevitable beginning of a new cycle.

Will this cycle take twenty billion *more* years, or a hundred? One can only speculate about the possibility. None the less, consider the following scenario. The energy that sustains the spreading macron we call the universe will inevitably drift into patterns that demand renewal. Things will slow to cessation, and, in a cosmic instant, the gradual wallowing process of reversal begins.

Throughout the universe a gentle, persistent, gravitational inhaling is born. All the patterned energy we call matter will sense the silent, invisible presence of its source of pattern. Some call it the center. The center of gravity. That center owns the periphery as surely as the periphery defines the center.

It is as these pages have claimed a locus of consciousness. Buckminster Fuller calls it love. Gravity is love. It is gravity that brings things together. Mysterious gravity . . . so shattering to the Western mind that our physicists claim it is born of particles that have no mass, no charge, and —so far—none have been found. Gravity is thus very much like love . . . unexplainable, undetectable by objective technologies, but so dominant everywhere.

The center of consciousness of the universe, expressing its love to its outward pulsing offspring, eventually calls them back. Each titanic galaxy, all the fiery and burned-out stars, all the billions of planets and asteroids . . . all the non-coalesced dust . . . even all the light responds to the call of the central consciousness. To the love. It is as though all the patterns silently *Nana I Ke Kumu* . . . silently look to the source.

It may take as long to breathe in as it did to breathe out, but when it does . . . the limits are reached. When all the matter, the patterned energy, comes together . . . a rebirth occurs. All the old patterns are lost. Patterns we cannot even conceive of are formed. The universe coalesces, and ways of being that still lie beyond speculation reform all substance. A cosmic child born of love begins its outward drift. The universe beings a new radiating, pulsating breath . . . again toward infinity.

Our way of being is locked—because we have chosen it so—to such a narrow time perspective that we see permanence only in our toys. Our buildings, books, monuments, and art. The pyramids, the Parthenon, and the Lincoln Memorial are but a flickering in time when compared to the age of the rocks from which they were made.

We have come to garner wealth manifested in the artifacts of culture. We are barren in our attempts to understand how long it takes a redwood tree to grow or a canyon to form, unless we can explain it against a war in history, the birth of a prophet, or the signing of a document in historic time.

We look to the cultural encrustations on our planet's surface for descriptions of our relationships to time and space. That cultural encrustation can be represented in the two hours it takes now to get from New York to Paris is sufficiently finite for it to become a cancer of the consciousness. We become diseased as we begin to make judgments about the extension of our encrustation. A pipeline here, a freeway there, a dam in that valley. Each one laying upon infinity. It did not take a million years to make the Grand Canyon. It took *twenty billion years*!

Our concept of time and its mirror image twin, space, has become so locked into the confines of culture and so distant from nature that, in a sense, it has become our enemy. In our trying to see nature whole, time and space disappear. There is no separating the events in the universe, locating them where they happen. After the next inward pulse, will we be led to conceive of time and space as starting at zero? Or will humans be tempted to postulate this as time-2/ space-2? How can we know whether or not it is time-2/ space-2 or possibly time-94/ space-94? Strangely, time and space as concepts may be among the most influential ways of thinking that create distance between culture and nature.

Russell Means, an Ogalala Sioux patriot, clearly encapsulates the Western or, as I would put it, the Northern world view. He says, in the December 1980 issue of *Mother Jones*:

> This is why 'truth' changes so fast in the European mind: the answers which result from such a process are only stop gaps, only temporary, and must be continuously discarded in favor of new stop gaps which support the mechanical models and keep them (the models) alive.[1]

Time-space figure strangely into another context . . . the difference between religion and spirituality. The differences between Northern and Southern hemisphere modes of consciousness (Chapter 13) contain within them totally different perceptions of time and space. The great world religions are a Northern hemisphere phenomenon. A nature-based spirituality characterizes the belief systems of the Southern hemisphere and the Native American vision. As citizens of the North, how may our view of time and space have affected a distinction between religion and spirituality?

Vine DeLoria, in *The Metaphysics of Modern Existence*, characterizes it this way:

> The great innovation of the world religions is to reduce natural events to a sequence containing some form of predictability, to introduce the conception of law and regularity into the natural world. Such an innovation is wholly artificial and *may be understood by primitive peoples as the original sin*. Certainly the acquisition of knowledge is understood in Genesis as the original sin, and it is ironic that in attempting to refine religious experiences into a more precise understanding that the great world religions commit the sin that alienates our species from the rest of the natural world (my italics).[2]

Religion is then an expression of the systematization of a belief system that designates being in terms of being reliable and predictable. It is the *logos* dualism again(Chapter13,pages172-178).Whereas science systematizes time and space to form knowledge, religion systematizes time and space to form theological doctrine.

Religion is linked to the words spirituality and spiritual only at the *third* definition level of most contemporary dictionaries. Yet in speaking to most persons who profess high religiosity, one will find spirituality being considered a synonym of religiosity. *These words are not synonymous as they have been used here.* In this book, spirituality has been defined as such:

Spirituality is that quality of being
that expresses the bonding
of all living
and non-living things
into an evolutionary unity.

"Spirituality" in the world religions and the Western world religions in particular expresses the sectarian bonding of humans to each other within a framework of a systematized, reliable, and predictable belief system. It is rare if not impossible to find spirituality as I have defined it expressed in a central way in the churches of the United States.Coupling this with the Old and New Testaments' vilification of wilderness, it appears normal that the places on the planet unconquered by technological culture afford fearful refuge. Fearful in that both testaments cite wilderness as the territory of devils and demons. For example, René Dubos in *The Wooing of Earth* cites, "After Jesus was baptized in the Jordan River, he was led up by the Spirit into the wilderness to be tempted by the devil."[3]

As surely as nature has been victimized by the philosophical premise of the major world religions, it has been even more abused by the technological application of these philosophies. Spirituality, from the premise of the world view of the major religions, *required* the effective separation of humans from nature. It displaced "spirituality" from nature to the culture. Tacitly, this displacement gave birth to the beliefs that one was doing good by turning the wilderness into a tamed and therefore more "spiritual" domain. The Northern world view in both science and religion gave birth and endowment to the planetary exploiter. In addition, in both science and religion those who did not make this separation were called "heathens" or "savages."

It may seem to some that the growth of attention toward the preservation of natural being and wilderness is the ghost wind of a possible reawakening. I believe it is. The Sierra Club, the Nature Conservancy, Friends of the Earth, Audubon, the National Wildlife Federation, and the American Wilderness Alliance are all expressions of commitment born of the middle class of the United States. The establishment. Others such as Greenpeace, a dozen antinuclear alliances, and the invisible network of more radical groups represent the possibility of an essentially anti-establishment wing of the reawakening.

To all of those, organized or not, who choose to participate in this transformation,

I would invite clarity on the differences I cite between religiosity and spirituality. In the attempt to reinstate planetary consciousness, most of us have been armed by the cultural perspectives of religiosity. We are taught to organize, take action, win-lose, and organize again. In other words, the religion of politics and technocratic vision may be the only commitment one has. This person *cannot* restore planetary vision.

The planet deserves more than the glibness and sloganeering of zealots. Zealots, hollow in their own being, always lead movements back into the dead embayments of their causes. I propose that in the past much of the frustration and anguish felt by environmentalists has resulted from *their being the same kind of persons they were attacking*. Being technocratic while attacking technocrats is hollow armor.

Far too many of my colleagues moved from war protests in the sixties and and seventies to environmental protests in the eighties. Some entered the "ministry" of the planet with the same zeal that proselytizers entered other ministries decades ago. They brought essentially a dependency/manipulative mode person—a Consciousness II commitment—to an ecology of transformation. I believe planetary consciousness requires far more holism. Salvation for the planet will require the methodologies and wisdom of the culture. It will require politics, technology, and religiosity. *But these are not enough!* Hollow humans create hollowness in their achievements.

Spirituality in the planetary sense must provide the balance for reflexive Western activism, both in the sense of science and religion. Natural consciousness must be brought into harmony with cultural consciousness. We cannot achieve harmony in consciousness via an exaggeration of adversity. We cannot thrash about looking for charismatic leaders as has been the pattern in the past. Each person is the leader, because each person is also the follower. There is no single way . . . recall the Tao—*the Tao that can be told is not the Tao*. In this is the germ plasm for the dissolution of zealotry. After all, zealotry is the "spiritual" methodology, of the fascist. The planet is not mother to the fascist . . . ego parents fascism.

We must go to the mother, the earth. We must make sure to talk with her, sleep in her, love her . . . This will give us the awesome immersion that rekindles in each of us the *being* that we are, the *being* that we have always been. Rolf Edberg in *The Dream of Kilimanjaro* writes:

> Can the cells remember? The cells in our skin and muscles, in the blood and nerve-endings, irises and eardrums? Remember over hundreds of thousands, millions of years?[4]

Another planetary voice . . . that of Peter Matthiessen, writing in *Parabola Winter 1980*, pursues this unity by claiming that traditional people the world over have much to teach our "spiritually crippled" people. Matthiessen quotes Lame Deer, who argues that whites "see with only one eye." Our half-blindness is claimed to be a curse upon us, but we were not always so. At one time, say the Native Indians, we too were in harmony. We too were in synchrony with the music of nature, with the great mystery. We can get it back, claims Matthiessen:

> As a first step we might consider this Great Mystery that is all about us . . . It is the music of the stars, the color of the wind, the dead stillness between tides . . . it is no less and no more strange . . . than our life itself.[5]

Spirituality is born of this re-established bonding with the cosmos. It is not a religion. Religions serve well to bond people to people . . . to align cultures with their human context gods. But spirituality as I have defined it earlier is more than bonding with human doctrine.

A friend, Willi Unsoeld, told the story of his first view of Everest, the mountain he had sworn to climb when he was a child. As he looked at the mountain, he first saw it as a human goal . . . an altar to be conquered at the end of a trail of discarded oxygen bottles, food packages, tents, and all forms of human debris. His mind at first was filled with logistics, schedules, routines, and material. He chanced to look down at his feet, where he noticed a small flower nearly crushed by his climbing boot. It was then and there he sensed complete unity. All the technology fell into the background. All the ritualistic routines of the militant side of climbing dissolved. Tears filled Willi's eyes. He, the flower, the mountain, and everything were all one.

"Nature is everywhere," Willi told me later. "It's in a vacant lot, a forest glen, in a rock . . . it's everywhere. You don't have to climb Everest to be with Mother Earth."

So our source is everywhere. We can bury it in cultural artifacts, but it inevitably pokes through. Each of us can go to a place to find reunion, and it may be moments away. Annie Dillard went to her backyard, and wrote the Pulitzer-prize winning testimony to *A Pilgrim at Tinker's Creek*.

Peter Matthiessen, who was quoted earlier, went to Nepal in September, 1973, to aid a biologist in a quest of the Himalayan Blue Sheep. Out of this journey was born a captivating book, *The Snow Leopard*. Immersing himself in the awesome texture of the Himalayas, Matthiessen documented his drifting between cultural and planetary consciousness. It was clear that nature made his perception of culture more poignant, and vice versa. His mind wandered freely, not only in the present, but also the past His own "reunion" with this kind of holonomic vision allowed him to repattern experiences years before. One such sensation, regarding his sudden synchrony with his son Alex, went as follows:

> In his first summers, forsaking all his toys, my son would stand rapt for near an hour in his sandbox in the orchard, as doves and redwings came and went on the warm wind, the leaves dancing, the clouds flying, birdsong and sweet smell of privet and rose. The child was not observing; *he was at rest in the very center of the universe, a part of things, unaware of endings and beginnings, still in unison with the primordial nature of creation, letting all light and phenomena pour through* (my italics).[6]

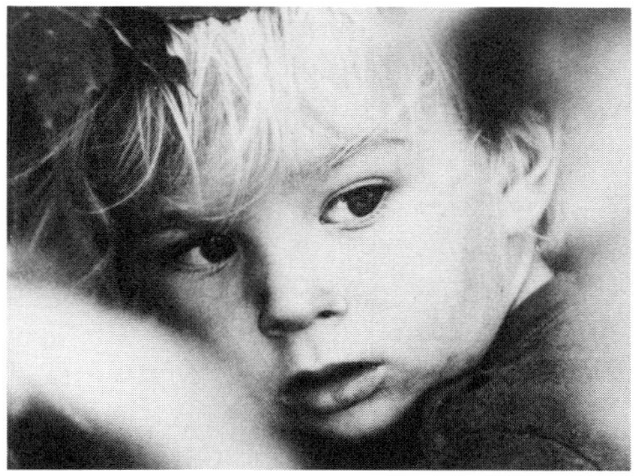

What Matthiessen describes is holonomy. Children are so good at it. They can immerse, blend, and transform their consciousness in all things. They do this because their patterns are yet so close to both their mothers. Both their human and planetary mothers gave them the dominion of virtual consciousness, oceanic consciousness, and the consciousness of the ontogenetic journey through all our phylogenetic ancestors. They have not forgotten. As such, they can revisit their origins.

But culture has in the past changed that. It invited children away from *Kumu* . . . the source. Yet this process seems to have the germ of change within it. If we honor the realities of evolving consciousness, in each individual, then we must tentatively suppose it will soon be manifested in the culture. Some have claimed the "human potential movement' testifies to this change. Others argue it is but an expression of organized narcissism. Again, hopefulness is confronted by cynicism.

Theodore Roszak, who is definitely a spokesperson for emerging consciousness, does not favor the notion of a human potential "movement." Roszak feels "movement" carries a connotation that is too political. He, like I, believes that the restoration of planetary consciousness is apolitical. Roszak prefers the connotation of a mosaic. The "human potential mosaic." Recall Vine Deloria's statement from Chapter 11 where he said, "changes in culture can be compared with the gradual change in the pieces of a mosaic." He warns us of the huge discomfort that accompanies such a process. The transformation of consciousness has no technological precedent. History documents how we have gone away from nature rather than how we can return to it.

What can one do then? Do I buy some turquoise jewelry, a membership in the Sierra Club, take Transcendental Meditation and eat organic food? Should I study Zen if I am an Anglo, or become born-again if I am an Asian? With what trappings may I adorn my consciousness? Consider the following:

In Canada recently I witnessed a sincere and compassionate effort on the part of educators to teach to the strengths of Native Indian children. It is well known that people who do not have written languages often show awesome visual acuity. These Native Indian children did. The Anglos recognized this and began to honor the children's visual expertise via guided imagery.

The results were remarkable. At the participation level, the children became excited, involved, and deeply expressive. But a deep sense of uneasiness was apparent during the class discussions. The teachers were so enthused at the students' quantum-jump performance that they overlooked or explained away the post-experience pall that settled over the group.

I flashed back on my own work with Navajo children a decade earlier. I was seeing an instant replay of my life. As a compassionate Westerner, I *could* review my own teaching biases and attempt to shift them and teach to a native child's strengths. Ten years ago in Crownpoint, New Mexico, I did the same thing to the Navajo children that these teachers did in Canada. I provided assignments that honored their visual abilities. Then I asked the students questions about their experience. I guided them toward further conclusions and elaborations about their interpretations. That is when the pall settled. Native Indian languages do not use the interrogative the way that European languages do. *Inquiry and interrogation is not a Native Indian methodology*. The aggression of questions so normal to our goal-directed biases was alien to them. For many Native Indian children, a question is an aggression upon their silence. So, regardless of how compassionate I was throughout the activity to this point, in the end I resorted to aggression. I forced the children to engage in an act that was unnatural to them. This act was imbedded in and approved by my Western, inquiry-based world view.

Edward Hall tells of an agricultural extension agent who one springtime began a program of demonstration plowing for some of the northern New Mexico pueblos. His message was to be helpful, but was absorbed by an instantly-generated atmosphere of disbelief. The northern pueblos were aghast, for this was the time when the earth had lost the winter hardness and her body was yielding and vulnerable. Slicing her body with a plow was a sacrilege equivalent to rape.

The Canadian teachers, me, the agricultural agent . . . We are all experts of purpose, of goal achievement, of getting things done. We are spokespersons for our technologies, rather than our beings. In the instances described, we were out for some gain. Russell Means says *being* is a spiritual act, gain is a *material* one. In nature, there is only being . . . never gain.

Human spirituality must eventually become identical with natural spirituality. If the organism affects the ecology, then the ecology affects the organism. It has been culture's way to emphasize the closed-system ecology it creates and to subdue the role of the individual to conforming subserviance. Culture strives to take the individual out of the process of evolution. We now possess the consciousness to know this is distorted. In nature, evolution takes place primarily because of changes in individuals. The individual eventually defines the species in the evolutionary process, for it is within the adaptation and transformation of the individual that the species persists.

What this means is clear. *Consciousness belongs to me!* I need not wait for cultural transformation to blend my being into planetary consciousness. I need no legislation to create a program to fund my being. I need no leader to unite me with the evolutionary flow of consciousness. In evolution, there are no leaders. The process is one of unity, thus each element of a unity is the entire unity.

In a way, some of the words used in this book may be poor metaphors. Could it be that transformation is really better said, *reunion*. Are we going to *higher* consciousness, implying stages and levels, or are we simply dissolving the concept of levels and, with grace, re-acquiring a sense of *unity* in consciousness? Culture has a way of giving us ladders when we need trees, reason when we need myth, and separateness when we need unity.

In the music of the universe, there is harmony. The discord, the nonharmonious, is slowly drifting back into the misty domains of our lost games. Ritual is being restored to rite. With a higher sense of the rhythms of the planet, we can recognize the emerging vision of grace. A grace to honor, not befowl . . . our mother. A grace to honor each other as end products of diverse cultural journeys. A grace to *become* the kind of human that can embody the spiritual. A grace to blend into all that is, was, and shall be.

When the last red man has vanished
from the earth,

And the memory is only the shadow
of a cloud moving across the prairie

These shores and forests
will still hold the spirits of my people

For they love this earth
as a newborn loves its mother's heartbeat.

If we sell you our land—
love it as we have loved it,

Care for it as we have cared for it.

Hold in your mind the memory of the land . . .

And with all your strength
With all your might
With all your heart

Preserve it for your children
and love it
as God loves us all.

Chief Sealth
Coastal Salish, 1854

ACKNOWLEDGEMENTS

All around me stand people who honor the planet. All of these are party to this book. Some friends whose names come easily to mind are ones who have given me both inspiration and insight into planetary consciousness. Bill Hammond, Larry Little Bird, Carol Sink, Annie Watson, Willi Unsoeld, Perl Charles, Tom Charles, Jim Graban, Jim Gladson, John Thompson, Nick Helburn, Bonnie McPherson, Bill and Suzanne Hering, Tom Crum, Laci Pook, David Randle, Ken Peterson, Jim McGrath, Bob Lepper, Robbie Porter, Mark Leary, Jake and Pat Nice, Ned Herrmann, Ric Cameron, Larry Petersen, Dave Kennedy, Tony Angell, Bob Silber, Dale and Linda Crider, Bob Hernbrode, Rodger Bybee, Bill Lucas, Milton McClaren, Dick Kienast, Don Davies.

Some have been explorers in these same regions of mind and being. They have risked writing, speaking, and acting as children of the planet. Jonas Salk, John Denver, O.J. Harvey, Charles Reich, Gregory Bateson, David Hawkins, David Galin, Ted Roszak, Frank Watson, Marilyn Ferguson, Jean Houston, Russell Means, Vine Deloria, Peter Matthiessen, Erich Jantsch, C.S. Brewster, George Gamov, Karl Pribram, Margaret Mead, Annie Dillard, William Irwin Thompson, George Leonard, Richard Jones, Dick Barnhart, Dick Gilbert, Gerry Kelly, Wallace Stegner, Bill Bridges, Dolores LaChapelle, Michael Hartoonian, Morris Udall, Jerome Bruner.

Some have moved in my dreams as light and warm winds do . . . Cheryl Charles, Sticie our son, Olina Gilbert, and Eve Thrall.

Some special mothers . . . Rosemarie Hammond, Ann Tucker, Irene Shelver, Diane Caven, Ann Denver, Pat Nice, Cathy Crum, Carolyn Roslund-Mercurio, Mattie Charles.

To Jan Rensel, Bardet Storyk, Lynn Flemming and Rocky Rohwedder who listened to many mutterings about the content in progress.

And Cheryl . . . who as editor, typist, retypist and grace maker, I cannot thank enough. This book is truly as much hers as mine.

And the Windstar family who share in commitment to the planet, Thomas F. Crum, Harold A. Thau, Steven Conger, Linda Conger, Terry Killam, Donna McIntyre, Karen Mason, Arthur Jackson, John Katzenberger, Keith Keating, Janet Britt, Greg Mozian, Mary Mann, Richard Kelm, Terrence Toole, Scott Mills, Judy Lowe, John Weiss, Carl Ellerbrook, John Edwards, Mardell Burkholder, Father Michael Abdo, David Armstrong, Bob Child, Ellis Deull, Nick DeWolf, R. Buckminster Fuller, Al Chung-Liang Huang, Robert Lewis, Antonietta L.F. Lilly, John C. Lilly, M.D., Amory Lovins, Mac McQuown, Stuart Mace, Milenko Matanovic, Alan Merson, W. Mitchell, Jim Morton, "Shorty" Pabst, Sally Ann Ranney, Roger Sant, Vicki Sant, David Sellers, Vince Shively, Michael Slosberg, David Spangler, Irwin M. Stelzer, Paul Winter. I thank and express love for being a part of what we do . . . .

This book was also born of special places on the planet. These too helped in the writing. Among them are Mt. Evans in mid-winter, the north coast of Kauaii, Hanauma Bay on Oahu, Arroyo de los Frijoles on the Pescadero Coast, the Sonoran desert in Arizona and Baja, almost anyplace in the plateau country of Colorado, Utah, and Arizona, the Indian Peaks Wilderness which I can see from the ridge behind Hawksong and a place called Windsong near the Colorado-Wyoming border.

A special thanks also goes to my editors at Addison Wesley who would surely be listed here except for policy. Their competence, patience, and generosity are unmatched. The freedom they extend to me in allowing me to create the visual and design dimension of my books provides a completeness and joy seldom experienced by authors.

The mind narrows when it is time to give thanks and honor sources, so I know the days ahead will bring many into my presence whom I forgot to cite here. But more than those I have erred in not mentioning, I will meet thousands more who should have been acknowledged.

It is from all who quest toward an honoring and gracing of the planet that this book is born. Thank you all.

Bob

FOOTNOTES

1: Mind, Beginning

[1]*The New Webster Encyclopedic Dictionary*. Chicago: Consolidated Book Publishers, 1971. page 180.

[2]Thomas, Lewis. *The Lives of a Cell*. New York: Bantam, 1975. page 2.

[3]Jantsch, Erich and Conrad H. Waddington. (Eds.) *Evolution and Consciousness*. Reading, Mass.: Addison-Wesley Publishing Co., 1976. page 37.

[4]Capra, Fritjof. *The Tao of Physics*. New York: Bantam, 1977. page 62.

2: Models of Mind

[1]Rose, Steven. *The Conscious Brain*. New York: Alfred Knopf, 1973. pp. 5-6.

[2]Ornstein, Robert. *The Psychology of Consciousness*. New York: Viking, 1972.

[3]Samples, Bob. *The Metaphoric Mind: A Celebration of Creative Consciousness*. Reading, Mass.: Addison-Wesley Publishing Company, 1976.

[4]Bruner, Jerome. *On Knowing: Essays for the Left Hand*. Cambridge: Harvard, 1962.

[5]Gazzaniga, Michael. *The Integrated Mind*. New York: Plenum, 1978.

[6]Cassirer, Ernst. *An Essay on Man*. New Haven: Yale University Press, 1944. page 20. Cited in Deloria, Vine. *The Metaphysics of Modern Existence*. New York: Harper and Row, 1979. page 50.

3: The Holonomic Mind

[1]Leonard, George. *The Silent Pulse*. New York: Dutton, 1978. page 81.

[2]Bohm, David. "The Enfolding-Unfolding Universe." An interview conducted by Renee Webber. *Revision Quarterly*. Summer/Fall, 1978. page 31.

[3]Ferguson, Marilyn. *The Brain/Mind Bulletin*. Los Angeles: Interface Press.

[4]Erich Jantsch, *Evolution and Consciousness*. Jantsch, Erich and Conrad Waddington. (Eds.) Reading, Mass.: Addison-Wesley Publishing Company, 1976. page 72.

[5]*Ibid.*, page 137.

[6]Bohm, *Opt. cit.*

4: Open Mind

Tart, Charles. (Ed.) *Altered States of Consciousness*. Garden City: Doubleday Anchor, 1972. page 1.

[2]Krippner, Stanley. (Personal communication).

[3]Brandwein, Paul. Address to Human Habitat Conference. December, 1980.

[4]Leonard, George. *The Silent Pulse*. New York: Dutton, 1978. page 19.

[5]*ibid.*, page 123.

5: Holonomy

[1]Capra, Fritjof. *The Tao of Physics*. New York: Bantam, 1977. page 198.

[2]Sheldrake, Rupert. Reported in *Brain Mind Bulletin*, Marilyn Ferguson, Ed. August 3, 1981. Volume 6, number 13, page 1.

[3]Dubos, Rene. *Wooing the Earth*. New York: Scribners, 1980. page 12.

[4]Matthiessen, Peter. Native Earth *Parabola, Earth and Spirit*. Volume VI, No. 1. Winter 1980. page 17.

6: Toward A Holonomic Culture

[1]Reich, Charles. *The Greening of America*. New York: Bantam, 1971. page 21.

[2]Lasch, Christopher. *The Culture of Narcissism: American Life in an Age of Diminishing Expectations*. New York: Norton, 1978.

7: Holonomy and the Person

[1]Harvey, O.J., and Catherine Felknor. "Parent-Child Relations as an Antecedent to Conceptual Functioning." Boulder: O.J. Harvey, 1969. pp. 4-6.

[2]Jantsch, Erich. *The Self Organizing Universe*. Oxford: Pergamon, 1980. page 75.

[3]Roszak, Theodore, *Person/Planet*. New York: Doubleday, 1978. page xxiv.

8: Evolution

[1]Ferguson, Marilyn. *The Aquarian Conspiracy*. Los Angeles: J.P. Tarcher, 1980.

[2]Jantsch, *op. cit.*

[3]Prigogine, Illya. in Jantsch, Erich *The Self Organizing Universe*. Oxford: Pergamon Press, 1980. page 35.

9: The Way of Life

[1]Salk, Jonas. *The Survival of the Wisest*. New York: Harper and Row, 1973. page 29.

[2]Jantsch, Erich. *op. cit.* page 10.

10: Toward a Blending

[1]Spieler, Joseph. "An Interview with Peter Schwartz: After the Recession." *Quest Magazine*. September, 1980. pp. 26-33.

[2]Harmon, Willis. Personal communication. 2

[3]Revel, Jean-Francois. *Without Marx or Jesus*. Garden City: Doubleday, 1970.

11: Whisperings Within

[1]Deloria, Vine. *The Metaphysics of Modern Existence*. New York: Harper and Row, 1979. p. 8.

12: Mind Evolving

[1]Jantsch, Erich. Evolution: Self Realization through Self Transcendence. Jantsch, Erich and Conrad Waddington. *Evolution and Consciousness*. Reading, Mass.: Addison-Wesley, 1976. page 41.

[2]Samples, Bob. *The Metaphoric Mind: A Celebration of Creative Consciousness*. Reading, Mass.: Addison-Wesley Publishing Company, 1976.

[3]Yankelovic, Daniel. *New Rules*. New York: Random House, 1981.

[4]Stavrianos, Leften. *The Promise of the Coming Dark Age*. San Francisco: Freeman, 1976. p. i.

13: East/West: North/South

[1]Tsu, Lao. *Tao Te Ching*. Gia-Fu Feng and Jane English. Trans. New York: Random House, 1972. page 1.

[2]Capra, Fritjof. *The Tao of Physics*. New York: Bantam, 1977. page 62.

[3]Dubos, Rene. *The Wooing of Earth*. New York: Charles Scribner and Sons, 1980. page 10.

[4]Deloria, Vine. *The Metaphysics of Modern Existence*. San Francisco: Harper and Row, 1979. page 151.

[5]*Ibid.*

[6]Bridges, William E. *Spokesmen for the Self*. Scranton: Chandler Publishing, 1971. page 47.

14: Nana I Ke Kumu

[1]Means, Russell. Fighting Words on the Future of the Earth *Mother Jones News*. Vol. V., No. X. December, 1980. page 26.

[2]Deloria, Vine. *Op. cit*. page 154.

[3]Dubos, Rene. *Op. cit*. page 10.

[4]Edberg, Rolf. *The Dream of Kilimanjaro*. New York: Pantheon, 1976. page 10.

[5]Matthiessen, Peter. *Op. cit*. page 17.

[6]Matthiessen, Peter. *The Snow Leopard*. New York: Bantam, 1978. page 41.

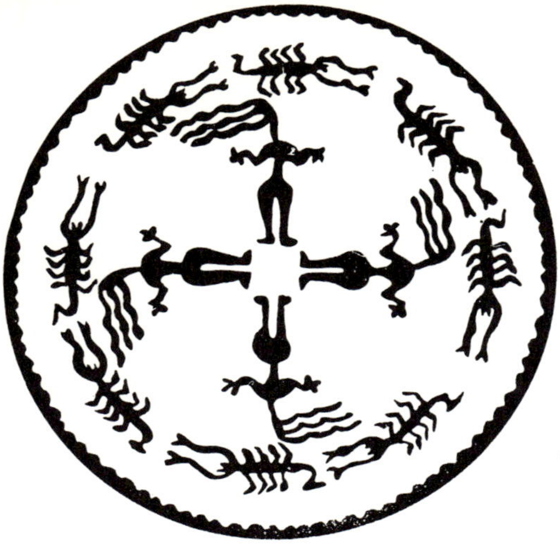

BIBLIOGRAPHY

The books cited in the footnote section were clearly the central books used in the gestation of this manuscript. However, others helped as well. This bibliography represents a partial list of allies.

Arguelles, Jose A. *The Transformative Vision*. Berkeley: Shambhala, 1975.

Bateson, Gregory. *Mind and Nature*. New York: E.P. Dutton, 1979.

Berry, Wendell. *The Unsettling of America: Culture and Agriculture*. New York: Avon, 1977.

Boddy, John. *Brain Systems and Psychological Concepts*. New York: John Wiley & Sons, 1978.

Brown, Barbara B. *Supermind*. New York: Harper & Row, 1980.

Brown, Harrison. *The Human Future Revisited*. New York: W.W. Norton, 1978.

Buzan, Tony and Terence Dixon. *The Evolving Brain*. New York: Holt, Rhinehart & Winston.

Coles, Robert. *Eskimos, Chicanos, Indians*. Volume IV of *Children of Crisis*. Boston: Little, Brown & Company, 1977.

Collier, John. *On the Gleaming Way*. Chicago: Sage Books, 1962.

Dawkins, Richard. *The Selfish Gene*. New York: Oxford University Press, 1976.

Dillard, Annie. *Pilgrim at Tinker Creek*. New York: Harper's Magazine Press, 1974.

Dychtwald, Ken. *Bodymind*. New York: Pantheon Books, 1977.

Farb, Peter. *Humankind*. New York: Bantam, 1980.

Fowles, John and Frank Horvat. *The Tree*. Boston: Little, Brown & Company, 1979.

Fuller, R. Buckminster. *Critical Path*. New York: St. Martin's Press, 1981.

Fuller, R. Buckminster, *Synergetics 2*. New York: Macmillan, 1979.

Furst, Charles. *Origins of the Mind*. Englewood Cliffs: Prentice-Hall, 1979.

Gardner, Howard. *The Shattered Mind*. New York: Alfred A. Knopf, 1975.

Goleman, Daniel and Richard J. Davidson, Editors. *Consciousness*. New York: Harper & Row, 1979.
Hall, Edward T. *Beyond Culture*. Garden City: Anchor Press/Doubleday, 1976.
Hampden-Turner, Charles. *Maps of the Mind*. New York: Macmillan, 1981.
Koestler, Arthur. *Janus*. New York: Random House, 1978.
LaChapelle, Dolores. *Earth Wisdom*. Los Angeles: The Guild of Tutors Press, 1978.
Lame Deer and Richard Erdoes. *Lame Deer, Seeker of Visions*. New York: Simon and Schuster, 1972.
Lee, Philip R. et al. *Symposium on Consciousness*. New York: Viking, 1976.
Lilly, John C. *Lilly on Dolphins*. Garden City: Anchor Press/Doubleday, 1975.
LeShan, Lawrence. *Alternate Realities*. New York: M. Evans & Company, 1976.
LeShan, Lawrence. *The Medium, the Mystic, and the Physicist*. New York: Viking, 1974.
Levi-Strauss, Claude. *The Savage Mind*. Chicago: University of Chicago Press, 1966.
Marschack, Alexander. *The Roots of Civilization*. New York: McGraw-Hill, 1972.
Mead, Margaret and Rhoda Metraux. *Aspects of the Present*. New York: William Morrow & Company, 1980.
Mead, Margaret. *Some Personal Views*. New York: Walker & Company, 1979.
Montagu, Ashley, Editor.*Learning Non-Aggression*.New York: Oxford University Press, 1978.
Montagu, Ashley.*The Nature of Human Aggression*.New York: Oxford University Press, 1976.
Nance, John. *The Gentle Tasaday*. New York: Harcourt Brace Jovanovich, 1975.
Neihardt, John G. *Black Elk Speaks*. New York: Pocket Books, 1972.
Ornstein, Robert E. *The Psychology of Consciousness*. New York: Viking, 1972.
Pelletier, Kenneth R. *Toward a Science of Consciousness*. New York: Delta Books, 1978.
Penfield, Wilder. *The Mystery of the Mind*. Princeton: Princeton University Press, 1975.
Pietsch, Paul. *Shufflebrain*. Boston: Houghton Mifflin, 1981.
Richards, Fred and Anne Cohen Richards. *Homonovus*. Boulder: Shields Publishing, 1973.
Sagan, Carl. *The Dragons of Eden*. New York: Random House, 1977.
Stegner, Wallace. *The Sound of Mountain Water*. New York: E.P. Dutton, 1980.

BIBLIOGRAPHY

Stone, Christopher D. *Should Trees Have Standing?* Los Altos: William Kaufmann, 1974.

Stone, Merlin. *When God Was a Woman*. New York: Harvest/Harcourt Brace Jovanovich, 1976.

Targ, Russell and Harold Puthoff. *Mind-Reach*. New York: Delacorte Press/ Eleanor Friede, 1977.

Thomas, Lewis. *The Medusa and the Snail*. New York: Viking, 1979.

Thompson, William Irwin. *The Time Falling Bodies Take to Light*. New York: St. Martin's Press, 1981.

Villoldo, Alberto and Ken Dychtwald, Editors. *Millennium*. Los Angeles: J.P. Tarcher, 1981.

Watts, Alan. *Tao: The Watercourse Way*.New York: Pantheon Books, 1975.

White, John, Editor. *Frontiers of Consciousness*. New York: Julian Press, 1974.

Wittrock, M.C., et al. *The Human Brain*. Englewood Cliffs: Prentice-Hall, 1977.

Worster, Donald. *Nature's Economy*. Garden City: Anchor Press/Doubleday, 1979.

Young, J.Z. *Programs of the Brain*. New York: Oxford University Press, 1979.

Zukav, Gary. *The Dancing Wu Li Masters*. New York: Bantam, 1979.

SOURCES

While this book is not intended to be a sourcebook as such, the following list represents a few organizations, journals and sources of inspiration that embody many facets contained in *Mind of Our Mother.*

Association for Humanistic Psychology
325 9th Street
San Francisco, California 94103

Association for Transpersonal Psychology
P.O. Box 3049
Stanford, California 94103

Brain Mind Bulletin
P.O. Box 42211
Los Angeles, California 90042

Dromemon: A Journal of New Ways of Being
G.P.O. Box 2244
New York, New York 10001

Human Dolphin Foundation
P.O. Box 4172
Malibu, California 90265

Newsletter of the Institute of Noetic Science
600 Stockton Street
San Francisco, California 94108

Parabola: A Journal
150 5th Avenue
New York, New York 10011

Revision Quarterly
P.O. Box 316
Cambridge, Massachusetts 02138

Solstice Seminars
Hawksong
Salina Star Route
Boulder, Colorado 80302

Windstar Foundation
Box 286
Snowmass, Colorado 81654